THE NEW FOREST

ITS HISTORY AND ITS SCENERY

BY

JOHN RICHARD
DE CAPEL WISE

With 63 illustrations. Drawn by Walter Crane.
Engraved by W. J. Linton.

And Two Maps.

British Library Cataloguing-in-Publication Data
A catalogue record for this book is available from the
British Library

CONTENTS

THE NEW FOREST FROM BRAMBLE HILL (SUNRISE).

PREFACE.

Under the title of the New Forest I have thought it best to include the whole district lying between the Southampton Water and the Avon, which, in the beginning of Edward I.'s reign, formed its boundaries. To have restricted myself to its present limits would have deprived the reader of all the scenery along the coast, and that contrast which a Forest requires to bring out all its beauties.

The maps are drawn from those of the Ordnance Survey, reduced to the scale of half an inch to the mile, with the additions of the names of the woods taken from the Government Map of the Forest, and my own notes.

The illustrations have been made upon the principle that they shall represent the scene as it looked at the time it was taken. Nothing has since been added, nothing left out. The views appear as they were on the day they were drawn. Two exceptions occur. The ugly modern windows of Calshot Castle, and the clock-face on the tower of the Priory Church of Christchurch, have been omitted.

Further, the views have been chosen rather to show the less-known beauties of the Forest than the more-known scenes. For this reason the avenue between Brockenhurst and Lyndhurst—the village of Minestead, nestling half amongst the Forest oaks and half in its own orchards—the view from Stoney Cross, stretching over wood and vale to the Wiltshire downs, have been omitted. Every one who comes to the Forest must see these, and every one with the least love for Nature must feel their beauty.

In their places are given the quiet scenes in the heart of the great woods, where few people have the leisure, and some not the strength to go—quiet brooks flowing down deep valleys, and woodland paths trod only by the cattle and the Forest workmen.

For the same reason, sunrise, and not sunset, has been chosen for the frontispiece.

To the kind help of friends I am indebted for much special aid and information—to the deputy surveyor, L. H. Cumberbatch, Esq., for permission to open various barrows and banks, for the use of the Government maps, as also for the Forest statistics—to the Rev. H. M. Wilkinson, and T. B. Rake, Esq., for great assistance in the botany and ornithology of the district; as also to Mr. Baker, of Brockenhurst, for the list of the Forest Lepidoptera.

London, November, 1862.

"Game of hondes he loued y nou, and of wÿlde best,
And hÿs forest and hÿs wodes, and mest þe nÿwe forest."

Robert of Gloucester's Chronicle
(concerning William the Conqueror)
Ed: Hearne, vol. ii. p. 375.

CHAPTER I.

INTRODUCTORY.

View in Bushey Bratley.

No person, I suppose, would now give any attention to, much less approve of, Lord Burleigh's advice to his son—"Not to pass the Alps." We have, on the contrary, in these days gone into an opposite extreme. We race off to explore the Rhine before we know the Thames. We have Alpine clubs, and Norway fishing, and Iceland exploring societies, but most of us are beyond measure ignorant of our own hills and valleys. Every inch of

Mont Blanc has been traversed by Englishmen, but who dreams of exploring the Cotswolds; or how many can tell in what county are the Seven Springs, and their purple anemonies? We rush to and fro, looking at everything, and remembering nothing. We see places only that we may be seen there, or else be known to have seen them.

Yet to Englishmen, surely the scenes of their own land should possess a greater interest than any other. Go where we will throughout England, there is no spot which is not bound up with our history. Nameless barrows, ruined castles, battlefields now reaped by the sickle instead of the sword, all proclaim the changes our country has undergone. Each invasion which we have suffered, each revolution through which we have passed, are written down for us in unmistakeable characters. The phases of our Religion, the rise or fall of our Art, are alike told us by the grey mouldings and arches of the humblest parish Church as by our Abbeys and Cathedrals. The faces, too, and gait, and dialect, and accent, of our peasantry declare to us our common ancestry from Kelt, and Old-English, and Norseman. A whole history lies hid in the name of some obscure village.

I am not, for one moment, decrying travelling elsewhere. All I say is, that those who do not know their own country, can know nothing rightly of any other. To understand the scenery of our neighbours, we must first see something of the beauties of our own; so that when we are abroad, we may be able to make some comparisons,—to carry with us, when we look upon the valleys of the Seine and the Rhine, some impression of our own landscapes and our own rivers—some recollection of our own cathedrals, when we stand by those of Milan and Rouen.

The New Forest is, perhaps, as good an example as could be wished of what has been said of English scenery, and its connection with our history. It remains after some eight hundred years still the New Forest. True, its boundaries are smaller, but the main features are the same as on the day when first afforested by the Conqueror. The names of its woods and streams and plains are the same. It is almost the last, too, of the old forests with which England was formerly so densely clothed. Charnwood is now without its trees: Wychwood is enclosed: the great Forest of Arden—Shakspeare's Arden—is no more, and Sherwood is only known by the fame of Robin Hood. But the New Forest still stands full of old associations with, and memories of, the past. To the historian it tells of the Forest Laws, and the death of one of the worst, and the weakness of the most foolish, of English kings. To the ecclesiologist it can show, close to it, the Priory Church of Christchurch, with all its glories of Norman architecture, built by the Red King's evil counsellor, Flambard; and just outside, too, its boundaries, the Conventual Church of Romsey, with its lovely Romanesque triforium, in whose nunnery Edith, beloved by the English, their "good Maud," "beatissima regina," as the Chroniclers love to call her, was educated.

At its feet lies Southampton, with its Late Norman arcaded town-wall, and gates, and God's House, with memories of Sir Bevis and his wife Josyan the Brighte, and his horse Arundel — the port for the Roman triremes, and afterwards for the galleys of Venice and Bayonne—where our own Henry V. built

> "the grete dromons,
> The Trinité, the Grace-Dieu."[1]

Within it, once in the very heart, stand the Abbot's house and the cloister walls of Beaulieu, the one abbey, with the exception of Hales-Owen, in Shropshire, founded by John. It can point, too, to the traces of Norman castles as at Malwood, to their ruins as at Christchurch, to Henry VIII. forts at Hurst and Calshot, built with the stones of the ruined monastery of Beaulieu; can show, too, bosomed amongst its trees, quiet village churches, most of them Norman and Early English, old manor-houses, as at Ellingham, famous in story, grey roadside crosses, sites of Roman potteries, and Keltic and West-Saxon battlefields and barrows scattered over its plains.

For the ornithologist its woods, and rivers, and seaboard attract more birds than most counties. For the geologist the Middle-Eocene beds are always open in the Hordle and Barton Cliffs inlaid with shell and bone. For the botanist and entomologist, its marshes, moors, and woodlands, possess equal treasures.

But in its wild scenery lies its greatest charm. From every hill-top gleam the blue waters of the English Channel, broken in the foreground by the long line of the Isle of Wight downs and the white chalk walls of the Needles. Nowhere, in extent at least, spread such stretches of heath and moor, golden in the spring with the blaze of furze, and in the autumn purple with heather, and bronzed with the fading fern. Nowhere in England rise such oak-woods, their boughs rimed with the frostwork of lichens, and dark beech-groves with their floor of red brown leaves, on which the branches weave their own warp and woof of light and shade.

Especially to its scenery I would call attention. This, above all, I wish to impress on the reader, seeing that beauty is one of the chief ends and aims of nature: and that the ground beneath our feet is decked with flowers, and the sky above our heads is painted with a thousand colours, to cheer man as he goes to his work in the morning, and to fill his heart with thankfulness as he returns at evening.

Now, neither are scarcely ever seen. The flowers cannot grow in our stony streets: the glory of the morning and evening is blotted out by the fog of smoke which broods over our cities.

As the population grows, our commons and waste lands disappear. Our large towns have swollen into provinces. Fashion sways the rich, Necessity compels the poor to live in them. As our wealth increases, our love for nature contracts. One, therefore, of the chief objects of this book is to show how much quiet beauty and how much interest lie beside our doors,—to point out to the reader who may be jaded by the toils of Fashion or Labour where in England there are still some thirty miles of moorland and woodland left uncultivated, over which he can wander as he pleases.

And here, if this book should induce any readers to visit the Forest, let me earnestly advise them to do so, as far as possible, on foot. I see but this main difference between rich and poor—that the poor work to get money, the rich spend money to get work. And I know no better way for Englishmen to use their superfluous energies than in learning their own country by walking over its best scenes.

I will only ask any one to make the experiment between walking and driving over the same ground; and see how much he will learn by the one, how much lose by the other method. In the one case, he simply hurries or stops at the discretion of some ignorant driver, who regards him of less importance than his horses; in the other, he can pause to sketch many a scene before invisible, can at his leisure search each heath or quarry for flowers or fossils, can turn aside across the field-paths to any village church, or wander through any wood which may invite him to its solitude, and, above all, know the pleasure of being tired, and the sweetness of rest in the noontide shade.

The Entrance from Barrow's Moor to Mark Ash.

Footnotes

1 *Political Pieces and Songs relating to English History.* Edited by Thomas Wright. Vol. ii., p. 199.

CHAPTER II.

ITS SCENERY.

The Stream in the Queen's Bower Wood.

As I said in the last chapter, one of the main objects of this book is to dwell upon the beauty of the Forest scenery. I chose the New Forest as a subject, because, although in some points it may not be more beautiful than many other parts of England—and I am glad to think so,—it gives, more than any other place, a far greater range of subject, in sea, and moor, and valley; because too, the traveller can here go where he pleases, without any of those lets and hindrances which take away so much pleasure; and, lastly, because here can best be seen Nature's crown of glory—her woods.

And, first, for a few words of general bearing upon this point. I do not think we ever estimate the woods highly enough, ever know their real worth, until we find some favourite retreat levelled to the ground, and then feel the void and irreparable blankness which is left. Consider, too, the use which Nature makes of her woods, either softening the horrors of the precipice, or adorning spaces which else would be utterly without interest, or adding beauty to beauty. Consider, further, how she beguiles us when we are in them, leading us forward, each little rise appearing a hill, because we cannot see its full extent; how, too, the paths close behind us, shutting us out with their silent doorways from all noise and turmoil, whilst the soft green light fills every dim recess, and deepens each pillared aisle, the floor paved with the golden mosaic of the sunlight.

For all these things is it that the woods have been, since the beginning of the world, the haunt of the flowers, the home of the birds, and the temple of man. The haunt of the flowers, I say, for in the early spring, before the grass is yet green in the meadows, here they all flock—white wood-anemones, sweet primroses, sweeter violets, and hyacinths encircling each stem with their blue wreaths. The home of the birds; for when the leaves at last have come, each tree is filled with song, and the underwood with the first faint chirping of the nestlings learning their earliest notes. As a temple for man, have they not been so since the world began? Taught by their tender beauty, and subdued by their solemn gloom, the imaginative Greek well consecrated each grove and wood to some Divinity. The early Christians fled to "the armour of the house of the Forest," to escape to peace and quietness. Here the old Gothic builders first learnt how to rear their vaulted arches, and to wreathe their pillars with stone arabesques of leaves and flowers, in faint imitation of a beauty they might feel, but never reach.[1]

Consider, too, the loveliness of all tree-forms, from the birch and weeping-willow, which never know the slightest formality, even when in winter barest of leaves, to the oak with its sinewy boughs, strained and tortured as they are in this very Forest, as nowhere else in England, by the Channel winds.[2] Consider, too, Nature's own love and tenderness for her trees,—how, when they have grown old and are going to decay, she clothes them with fresh beauty, hides their deformities with a soft green veil of moss and the grey dyes of lichens, and, not even content with this, makes them the support for still greater loveliness—drapes them with masses of ivy, and hangs upon them the tresses of the woodbine, loading them to the end of their days with sweetness and beauty.

All this, and far more than this, you may see in the commonest woods round Lyndhurst, in Sloden, in Mark Ash, or Bratley.

Then, too, there is that perpetual change which is ever going on, every shower and gleam of sunshine tinting the trees with colour from the tender tones of April and May, through the deep green of June, to the russet-red of autumn. Each season ever joins in this sweet conspiracy to oppress the woods with loveliness.

Taking a more special view, and looking at the district itself, we must remember that it is situated on the Upper and Middle Eocene, and presents all the best features of the Tertiary formation. Its hills may not be high, but they nowhere sink into tameness, whilst round Fordingbridge, and Goreley, and Godshill, they resemble, in degree, with their treeless, rounded forms, shaggy with heath and the rough sedge of the fern, parts of the half-mountainous scenery near the Fifeshire Lomonds.[3]

On the sea-coast near Milton, rise high gravel-capped cliffs, with a basis of Barton clays, cleft by deep ravines, locally known as "bunnies." Inland, valleys open out, dipping between low hills, whilst masses of beech and oak darken the plains. Here and there, marking the swamps, wave white patches of cotton-grass, whilst round them, on the uplands, spread long, unbroken stretches of purple heather; and wide spaces of fern, an English Brabant, studded with hollies and yews, some of them as old as the Conquest. Here and there, too, as at Fritham, small farmsteads show their scanty crops of corn, or, as at Alum Green and Queen's North, green lawns pierce and separate the woods, pastured by herds of cattle, with forest pools white with buckbean, and the little milkwort waving its blue heath on the banks.

These are the main characteristics of the New Forest, and, in some points at least, were the same in the days of the Red King. Nature, when left to herself, even in the course of centuries, changes little. The wild boars, and the wolves, and the red deer, are gone. But much else is the same. The sites and the names of the Forest manors and villages, with slight alterations, remain unchanged. The same barrows still uplift their rounded forms on the plains; the same banks, the same entrenchments, near which, in turn, lived Kelt, and Roman and Old-English, still run across the hills and valleys. The same churches rear their towers, and the mills still stand by the same streams.

The peasants, too, still value the woods, as they did in the Conqueror's time, for the crop of mast and acorns,—still peel off the Forest turf, and cure their bacon by its smoke.[4] The charcoal-burner still builds the same round ovens as in the days of William the Red. Old-English words, to be heard nowhere else, are daily spoken. The last of the old Forest law-courts is held every forty days at Lyndhurst. The bee-master—*beoccorl*—still tends his hives, and brews the Old-English mead, and lives by the labours of his bees. The honey-buzzard still makes her nest in the beeches round Lyndhurst, and the hen-harrier on the moors near Bratley.

I suppose this is what strikes most persons when they first come into the New Forest,—a sense that amidst all the change which is going forward, here is one place which is little altered. This is what gives it its greatest charm,—the beauty of wildness and desolateness, broken by glimpses of cultivated fields, and the smoke of unseen homesteads among the woods.

Yet the feeling is not quite true. Like every other place in England, it has suffered some change, and moved with the times. Instead of the twang of the archer's bow, the sunset gun at Portsmouth sounds every evening. The South-Western Railway runs through the heart of it; and in place of the curfew's knell, the steam whistle shrieks through its woods.

We do not see the forest of our forefathers. Go back eight centuries, and look at the sights which the Normans must have beheld,—dense underwoods of hollies on which the red deer browsed; masses of beech and chestnut, the haunts of the wolf and the boar; plains over which flocks of bustards half-ran, half-flew; swamps where the crane in the sedge laid its buff and crimson-streaked eggs; whilst above grey-headed kites swam in circles; and round the coast the sea-eagle slowly flapped its heavy bulk. Great oaks, shorn flat by the Channel winds, fringed the high Hordle cliffs, towering above the sea; and opposite, as to this day, rose the white chalk rocks of the Needles, and the Isle of Wight, not bare as it now is but covered, too, with a dense forest. And the sun would set as it now does, but upon all this further beauty, making a broad path of glory across the bay, till at last it sank down over the Priory Church of Christchurch, which Flambard was then building.

Gone, too, for ever all the scenes which they must have had of the Avon, glimpses of it caught among the trees as they galloped through the broad lawns, or under the sides of Godshill, and Castle Hill crested with yews and oaks.[5]

These may all be gone, but plenty of beauty is still left. The Avon still flows on with its floating gardens of flowers—lilies and arrowhead, and the loosestrife waving its crimson plumes among the green reeds. The Forest streams, too, still flow on the same, losing themselves in the woods, eddying round and round in the deep, dark, prison-pools of their own making, and then escaping over shallows and ledges of rolled pebbles, left dry in the summer, and on which the sunlight rests, and the shadows of the beeches play, but in the winter chafed by the torrent.

Across its broad expanse of moors the sun still sets the same in summer time—into some deep bank of clouds in the west, and as it plunges down, flashes of light run along their edges, and each thin band of vapour becomes a bar of fire, and the far-away Purbeck hills gleam with purple and amethyst.

The same sea, too, still heaves and tosses beneath the Hordle and Barton Cliffs, with the same purple patches, shadows of clouds, sailing over it, as its waves, along the shore, unroll their long scrolls of foam.

These great natural facts have not changed. Kelt and Roman have gone, but these are the same.

Nor must I forget the extraordinary lovely atmospheric effects, noticed also by Gilpin,[6] as seen under certain conditions, from the Barton Cliffs on the Isle of Wight and the Needles. Far out at sea will rise a low white fog-bank gradually stealing to the land, enveloping some stray ship in its folds, and then by degrees encircling the Island, whilst the chalk cliffs melt into clouds. On it still steals with its thick mist, quenching the Needles Light which has been lit, till the whole island is capped with fog, and neither sea nor sky are seen,—nothing but a dense haze blotting everything. Then suddenly the wind lifts the great cloud westward, and its black curtains drop away, revealing a sky of the deepest blue, barred with lines of light: and the whole bay suddenly shines out clear and glittering, the Island cliffs flashing with opal and emerald, and the ship once more glides out safe from the darkness.

A few words, too, must be said about the two principal tree-forms which now make the Forest. The oaks here do not grow so high or so large as in many other parts of England, but they are far finer in their outlines, hanging in the distance as if rather suspended in the air than growing from the earth, but nearer, as especially at Bramble Hill, twisting their long arms, and interlacing each other into a thick roof. Now and then they take to straggling ways, running out, as with the famous Knyghtwood Oak, into mere awkward forks. The most striking are not, perhaps, so much those in their prime, as the old ruined trees at Boldrewood, their bark furrowed with age, their timber quite decayed, now only braced together by the clamps of ivy to which they once gave support and strength.[7]

The beeches are even finer, and more characteristic, though here and there a tree sometimes resembles the oaks, as if with long living amongst them, it had learnt to grow like them. The finest beech-wood is that of Mark Ash. There you may see true beech-forms, the boles spangled with silver scales of lichen, and the roots—more fangs than roots—grasping the earth, feathered with the soft green down of moss.

But not in individual trees lies the beauty of the Forest, but in the masses of wood. There, in the long aisles, settles that depth of shade which no pencil can give, and that colouring which no canvas can retain, as the sunlight pierces through the green web of leaves, flinging, as it sets, a crown of gold round each tree-trunk.

Let no one, however, think they know anything of the Forest by simply keeping to the high road and the beaten tracks. They must go into it, across the fern and the heather, and, if necessary, over the swamps, into such old woods as Barrow's Moor, Mark Ash, Bushey Bratley, and Oakley, wandering at their own will among the trees. The best advice which I can give to see the Forest is to follow the course of one of its streams, to make it your friend and companion, and go whereever it goes. It will be sure to take you through the greenest valleys, and past the thickest woods, and under the largest trees. No step along with it is ever lost, for if never goes out of its way but in search of some fresh beauty.

We see plenty of pictures in our Exhibitions from Burnham Beeches or Epping Forest, but in the New Forest the artist will find not only woodland, but sea, and moorland, and river views. There are, as I have said, when taken in details, more beautiful spots in England, but none so characteristic. Finer trees, wilder moors, higher hills, more swiftly-flowing brooks, may be found, but nowhere that quietness so typical of English scenery, yet mixed with wildness, nowhere so much combined in one.

I say, too, this, strange as it may doubtless appear, that Government, whenever it fells any timber, should spare some of the finest trees for the sake of their beauty, and for the delight they will give to future generations. Cut down, and sawn into planks, they are worth but so many pounds. Standing, their value is inappreciable. We have Government Schools of Design, and Government Picture Galleries, but they are useless without Nature to assist the student. Government, by keeping here some few old trees, will do more to foster true Art than all the grants of Parliament. The old thorns of Bratley, the beeches of Mark Ash, and the yews of Sloden, will teach more than all the schools and galleries in the world. As we have laws to preserve our partridges and pheasants, surely we might have some to protect our trees and our landscapes.

Lastly, from its very nature, the New Forest is ever beautiful, at every season of the year, even in the depth of winter. The colouring of summer is not more rich. Then the great masses of holly glisten with their brightest green; the purple light gathers round the bare oaks, and the yews stand out in their shrouds of black. Then the first budding branch of furze sparkles with gold, and the distant hill-side glows with the red layers of beech-leaves. And if a snow-storm passes up from the sea, then every bough is suddenly covered with a silver filigree of whitest moss.

This joyful tyranny of beauty is ever present, at all times and hours, changeful in form, but the same in essence. Year after year, day after day, it appears.

I know, however, it is impossible to make people see this beauty, which, after all, exists only in each beholder's mind. No two people see the same thing, and no person ever sees it twice. But, I believe, we may all gain some idea of the glory which each season brings—some glimpses of the heaven of beauty which ever surrounds us—if we will seek for them patiently and reverently. They cannot with some be learnt at once, but, in degrees, are attainable by all; but they are attainable only upon this one condition,—that we go to Nature with a docile, loving spirit, without which nothing can be learnt. If we go with any other feeling, we had much better stay in a town amidst the congenial smoke, than profane Nature with the pride of ignorance and the insolence of condescension.

The Charcoal-Burner's Path, Winding Shoot

Footnotes

1. It is worth noticing how, according to their natures, our English poets have dwelt upon the meaning of the woods, from Spenser, with his allegories, to the ballad-singer, who saw them only as a preserve for deer. Shakspeare touches upon them with both that joyful gladness, peculiar to him, and the deep melancholiness, which they also inspire. Shelley and Keats, though in very different ways, both revel in the woods. To Wordsworth they are—
"like a dream
Or map of the whole world: thoughts, link by link,
Enter through ears and eyesight, with such gleam
Of all things that at last in fear I shrink."

Of course, under the names of woods, and any lessons from them, I speak only of such lowland woods as are known chiefly in England; not dense forests shutting out light and air, without flowers or song of birds, whose effect on national poetry and character is quite the reverse to that of the groves and woodlands of our own England. See what Mr. Ruskin has so well said on the subject. *Modern Painters*, vol. v., part vi, ch. ix., § 15, pp. 89, 90; and, also in the same volume, part vii., chap, iv., § 2, 3, pp. 137-39; and compare vol. iii., part iv., ch. xiv., § 33, pp. 217-19.

2. In the lower part of the Forest, near the Channel, the effect is quite painful, all the trees being strained away from the sea like Tennyson's thorn. It is the *usnea barbata* which covers them, especially the oaks, with its hoary fringe, and gives such a character to the whole Forest.

3. The reader must bear in mind that the word "forest" is here used, as it is always throughout the district, in its primitive sense—"foresta est tuta ferarum mansio," "statio ferarum." (See Dufresne on the word "foresta.") And the moors and plains are so called, though there may not be a single tree growing upon them.

4. The woods, in *Domesday*, are generally valued by the number of swine they maintain. Thus, under Brockenhurst we find "silva de 20 porcis;" that is, a wood capable of supporting twenty hogs. Curiously enough, there is no mention of charcoal-burning in the New Forest in *Domesday*, though we know, from other sources, that it was carried on to some extent.

5. For a justification of this general picture, I must refer the reader to the next chapter, where references to *Domesday*, as to the state of the district before its afforestation by the Conqueror, and the evidence supplied by the names of places, are given. I may add, as showing the former nature of the woods, that the charcoal found in the barrows, embankments, and the Roman potteries, is made from oak and beech, but principally from the latter. Since, too, the deer have been destroyed, the young shoots of

holly are springing up in every direction, and another generation may again see the Forest still more resembling its old condition. As a proof that the Hordle Cliffs were covered with timber, the fishermen dredging for the *septaria* in the Channel constantly drag up large boles of oaks, which are locally known as "mootes." The existence of the chestnut is shown by the large beams in some of the old Forest churches, as at Fawley; but none now exist, except a few, comparatively modern, though very fine, at Boldrewood. Further, the Forest could never, except in the winter, have been very swampy, as the gravelly formation of the greater part of the soil supplies it with a natural drainage. Still, there were swamps, and in the wet places large quantities of bog-oak have been dug up, bearing witness, as in other countries, of an epoch of oaks, which preceded the beech-woods. Gough, in his additions to Camden's *Britannia*, vol. i., p. 126, describes Godshill as being in his day covered with thick oaks. When, too, Lewis wrote in 1811, old people could then recollect it so densely covered with pollard oaks and hollies that the road was easily lost. (*Historical Enquiries on the New Forest*, p. 79, Foot-note.) No one, I suppose, now believes that wolves were extirpated by Edgar. They and wild boars are expressly mentioned in the Laws of Canute (*Manwood: a Treatise of the Lawes of the Forest*, f. 3, § 27, 1615), and lingered in the north of England till Henry VIII.'s reign. (See further on the subject, *The Zoology of Ancient Europe*, by Alfred Newton, p. 24.) I have hesitated, however, to include the beaver, though noticed by Harrison, who wrote in 1574, as in his time frequenting the Taf, in Wales (*Description of England*, prefixed to Holinshed's *Chronicle*, ch. iv. pp. 225, 226.) The eggs of cranes, bustards, and bitterns, were, we know, protected as late as the middle of the sixteenth century. (*Statutes of the Realm*, vol. iii., p. 445, 25° Henry VIII., ch. xi., § 4; and vol. iv., p. 109, 3°, 4°, Ed. VI., ch. vii.) The last bustard was seen in the Forest, some twenty-five years ago, on Butt's Plain, near Eyeworth. It is a sad pity that the enormous collection of birds' bones, described as chiefly those of herons and bitterns, found by Brander amongst the foundations of the Priory Church at Christchurch (see *Archæologia*, vol. iv, pp. 117, 118), were not preserved, as they might have yielded some interesting results. We must, however, still bear in mind that there are far more points of resemblance than of difference between the Forest of to-day and that of the Conqueror's time; especially in the long tracts of fern and heath and furze, which certainly then existed, pastured over by flocks of cattle.

6. *Remarks on Forest Scenery, illustrated by the New Forest*, vol. ii., pp. 241 -46; third edition. Some mention should here he made of Gilpin, a man who, in a barren, unnatural age, partook of much of the same spirit as Cowper and Thompson, and whose work should be placed side by side with their poems. Unfortunately, much of his description is now quite useless, as the Forest has been so much altered; but the real value of the book still remains unchanged in its pure love for Nature and its simple, unaffected tone. It is well worth, however, noticing—as showing the enormous difficulty of overcoming an established error—that, notwithstanding his true appreciation of bough-forms (see vol. i., pp. 110-12, same edition), and his hatred of pollarded shapes, and all formalism (some vol., p. 4), he had not sufficient force to break through the conventional drawing of the eighteenth and the beginning of the nineteenth centuries, and his trees (see, as before, pp. 252-54) are all drawn under the impression that they are a gigantic species of cabbage. The edition, however, published in 1834, and edited by Sir T. D. Lauder, is in this and many other respects, far better.

7. The following measurements may have, perhaps, an interest for some readers:—Girth of the Knyghtwood oak, 17 ft. 4 in.; of the Western oak at Boldrewood, 24 ft. 9 in.; the Eastern, 16 ft.; and the Northern, in the thickest part, 20 ft. 4 in.; though, lower down, only 14 ft. 8 in.; beech at Studley, 21 ft.; beech at Holmy

Ridge, 20 ft. The handsomest oak, however, in the district, stands a few yards outside the Forest boundary, close to Moyle's Court, measuring 18 ft. 8½ in.

CHAPTER III.

ITS EARLY HISTORY.

The Cattle Ford, Liney Hill Wood.

ONCE the New Forest occupied nearly the entire south-west angle of Hampshire, stretching, when at its largest, in the beginning of the reign of Edward I., from the Southampton Water on the east, whose waves, the legend says, reproved the courtiers of Canute, to the Avon, and even, here and there, across it, on the west; and on the north from the borders of Wiltshire to the English Channel.

These natural boundaries were, as we shall see, reduced in that same reign. Since then encroachments on all sides have still further lessened its limits; and it now stretches, here and there divided by manors and private property, on the north from the village of Bramshaw, beyond Stoney Cross, near where Rufus fell, or was supposed to fall, to Wootton on the south, some thirteen miles; and, still further, from Hardley on the east to Ringwood, the Rinwede of *Domesday*, on the west.

In the year 1079, just thirteen years after the battle of Hastings, William ordered its afforestation. From Turner and Lingard down to the latest compiler, our historians have represented the act as one of the worst pieces of cruelty ever committed by an English sovereign. Even Lappenberg calls the site of the Forest "the most thriving part of England," and says that William "mercilessly caused churches and villages to be burnt down within its circuit;"[1] and, in another place, speaks of the Conqueror's "bloody sacrifice," and "glaring cruelty towards the numerous inhabitants."[2] To such statements in ordinary writers we should pay no attention, but they assume a very different aspect when put forward, especially in so unqualified a way, by an historian to whom our respect and attention are due. I have no wish here to defend the character of William. He was one of those men whose wills are strong enough to execute the thoughts of their minds, ordained by necessity to rule others, holding firmly to the creed that success is the best apology for crime. Yet, too, he had noble qualities. Abroad he was feared by the bad, whilst at home such order prevailed throughout England, that a man might travel in safety "with his bosom full of gold" from one end to the other.[3]

What I do here protest against is the common practice of implicitly believing every tradition, of repeating every idle story which has been foisted into the text either by credulity or rancorous hatred—of, in fact, mistaking party feeling for history. The Chroniclers had every reason to malign William. His very position was enough. He had pressed with a heavy hand on the Old-English nobles, stripped them too often of their civil power, and their religious honours; and failing to learn, had, like a second Attila, tried to uproot their language.

The truth is, we are so swayed by our feelings that the most dispassionate writer is involuntarily biased. We in fact pervert truth without knowing we do so. Language, by its very nature, betrays us. No historian, with the least vividness of style, can copy from another without exaggeration. The misplacement of a single word, the insertion of a single epithet, gives a different colour and tone. And, in this very matter of the New Forest, we need only take the various accounts, as they have come down, to find in them the evidences of their own untruth.[4]

I do not here enter into the question of William's right to make the Forest—about this there can be no doubt—but simply into the methods which he employed in its formation.[5] The earliest Chronicler of the event, Gulielmus Gemeticensis, who has been so often quoted in evidence of William's cruelty, both because he was a Norman, and chaplain to the King, really proves nothing. In the first place, the monk of Jumieges did not write this account, but some successor, so that the argument drawn from the writer's position falls to the ground.[6] In the second place, his successor's words are—"Many, however, say (*ferunt autem multi*) that the deaths of Rufus and his brother were a judgment from heaven, because their father had destroyed many villages and churches in enlarging (*amplificandam*) the New Forest."[7] The writer offers no comment of his own, and simply passes over the matter, as not worth even refutation. His narrative, however, if it tells at all, tells against the common theory, as he states that William only extended the limits of a former chase.

The account of Florence of Worcester is, on the whole, equally unsatisfactory. His mention of the New Forest, like that, by the way, of most of the Chroniclers, does not occur in its proper place at the date it was made—when the wrong, we should have thought, must have been most felt—but is suggested by the death of Rufus, when popular superstition had come into play, and time had lent all the force of exaggeration to what must always have been an unpopular event. Florence,[8] however, speaks in general terms of men driven from their homes, of fields laid waste, and houses and churches destroyed; words, which as we shall see, carry their own contradiction. Vitalis,[9] too, not only declares that the district was thickly inhabited, but that it even regularly supplied the markets of Winchester, and that William laid in ruins no less than sixty parishes. Walter Mapes,[10] who flourished about the middle of the twelfth century, adds further that thirty-six mother churches were destroyed, but falls into the error of making Rufus the author of the Forest, which of course materially affects his evidence.

Knyghton,[11] however, who lived in the reign of Richard II., is doubtful whether the number of churches destroyed was twenty-two or fifty-two, an amount of difference so large that we might also reasonably suspect his narrative, whilst he also commits the mistake of attributing the formation of the Forest to Rufus.

Now, the first thing which strikes us is that as the writers are more distant in point of time, and therefore less capable of knowing, they singularly enough become more precise and specific. What Florence of Worcester speaks of in merely general terms, Vitalis, and Walter Mapes, and Knyghton, give in minute details down to the very number of the parishes and churches.[12]

As far as mere written testimony goes, we have nothing to set against their evidence, except *Domesday*, and the negative proof of *The Chronicle*. Not one word does The Chronicler, who, be it remembered, personally knew the Conqueror[13]—who has related each minute event of his reign, exposed each shortcoming, and branded each crime, say of the cruelty of the afforestation. Evidence like this, coming from such an authority, is in the highest degree important. The silence is most suggestive. It is impossible to believe, that so faithful an historian, had it been committed, should never have hinted at the devastation of so much property, and the double crime of cruelty and profanity in destroying alike the inhabitants and their churches.

But the briefest analysis of *Domesday*, and a comparison of its contents with those of the survey made in Edward the Confessor's reign, will more clearly show the nature and extent of the afforestation than any of the Chroniclers. From it we find that about two-thirds of the district, including some thirty manors, was entirely afforested. But it by no means carries out the account that the villages were destroyed and the inhabitants banished, or, according to others, murdered. For in some cases, as at Eling, it is noted that the houses are still standing and the inmates living in the King's Forest; and in others, as at Batramsley, Pilley, Wootton, and Oxley, express mention is made that only the woods are annexed, and that the meadows and pastures are not afforested, but remain in the hands of their former West-Saxon holders.[14] Again, too, we find that some of the manors, as at Hordle and Bashley, though considerably lessened, kept up their value. Others, as at Efford, actually doubled their former assessments. Still more remarkable, some again, as at Brockenhurst, Sway, and Eling, though reduced in size, increased one-third and two-thirds in value. One explanation can alone be given to such facts—that only the waste lands were enclosed, and the cultivated spared.

The village of Totton, though close to the Forest, was not touched, although all the neighbouring hamlets were in various degrees afforested, simply because it consisted of only pasture and plough-land, whose value had increased no less than one fourth. The hamlets of Barton—literally, Bere-tun, the corn village—and Chewton, where to this day is the best land in the south-west of Hampshire, were also spared; though we find all the neighbouring villages and manors, Milton, Beckley, Bashley, Fernhill, Whitefields, Arnwood, all more or less enclosed; the reason being, as was before said, that the Conqueror took only the waste lands and the woods.

In the woods which were afforested people were allowed to live; though, probably, they voluntarily left them, as labour could not there be so well obtained as in the unafforested parts.[16] In all other respects there seems to have been no disarrangement. Both on the outskirts and in the heart of the Forest, the villains and borderers still worked as before, carrying on their former occupations.[17] The mills at Bashley, and Milford, and Burgate, all in the Forest, went on the same. The fisheries at Holdenhurst and Dibden were undisturbed. The salterns at Eling and Hordle still continued at work, showing that the people still, as before, sowed and reaped their corn, and pastured and killed their cattle.

Again, in other ways, *Domesday* still more clearly contradicts the Chroniclers, as to the inhabitants being driven out of their homes. Canterton was held by Chenna of Edward, and still in *Domesday*, in part, remains in his possession. Ulviet, the huntsman, who had rented land at Ripley under Edward, still rents the same. His son, Cola, also a huntsman, holds land at Langley, which his father had held of Edward; whilst his other son, Alwin, holds land at Marchwood, which, also, his father had held. Saulf, a West-Saxon thane, who had held land at Durley of Edward, now holds it at Batramsley, and his wife at Hubborn.[18]

Ulgar, a West-Saxon, holds the fourth of a hyde at Milford, just as he had held it of Edward; with this difference, that it was now assessed at three-fourths of a rood, on account of the loss sustained by the woods being taken into the Forest. The sons of Godric Malf, another West-Saxon thane, hold the same lands which their father had held of Edward, at Ashley, Bisterne, Crow, and Minstead, the last property being rated at half its proper value because the woods were afforested. The West-Saxon Aluric rents property at Oxley, Efford, and Brockenhurst, which his father and uncle rented under Edward, and not only receives lands at Milford in exchange for some taken into the Forest, but actually buys estates at Whitefields from other West-Saxons.[19]

Such facts must be stronger than any mere history compiled by writers who were not only not near the spot, but the majority of whom lived a long time after the events they venture so minutely to describe.

But we have not yet exhausted the valuable evidence of *Domesday*. The land in the Forest district is rated at much less than in other parts of Hampshire, showing that it was therefore poorer, and not only the land, but the mills. Further—and this is of great importance, as so thoroughly overthrowing the common account—we find in that portion of the survey which comes under the title, "In Novâ Forestâ et circa cam," only two churches mentioned, one at Milford, and another at Brockenhurst, in the very heart of the Forest. Both stand to this hour, and prove plainly by their Norman work that William allowed them to remain.

Such is the evidence which *The Chronicle* and the short examination of *Domesday* yield. The country itself, however, still more plainly proves the bias of the Chroniclers. The slightest acquaintance with geology will show that the Forest was never fertile, as it must have been to have maintained the population which filled so many churches.[20] Nearly the whole of it is covered with sand, or capped with a thick bed of drift, with a surface-soil only a few inches deep, capable of naturally bearing little, except in a few places, besides heath and furze. On a geological map we can pretty accurately trace the limits of the Forest by the formation. Of course, in so large a space, there will be some spots, and some valleys, where the streams have left a richer glebe and a deeper tilth.[21]

But the Chroniclers, by their very exaggeration, have defeated their own purpose. There is in their narration an inconsistency, which, as we dwell upon it, becomes more apparent. We would simply ask, where are the ruins of any of the thirty or fifty churches, and the towns of the people who filled them? Why, too, did not the Chroniclers mention them specifically? Why, further, if William pulled down all the churches, are the only two, at Brockenhurst and Milford, recorded in *Domesday*,[22] still standing with their contemporary workmanship? Why, too, is Fawley church, with its Norman door, and pillars, and arches, formerly, as we know from another portion of *Domesday*, in the Forest, remaining, if all were destroyed? And why, last of all, if the inhabitants were exterminated, was a church built at Boldre, in the very wildest part of the Forest, immediately after the afforestation, and another at Hordle?[23]

Had there been any buildings destroyed, all ruins of them would not have been quite effaced, even in the course of eight centuries. The country has been undisturbed. Nature has not here, as in so many places, helped man in his work of destruction. They cannot, we know, have been built on, or ploughed over, or silted with sand, or choked with mud, or washed away by water. The slightest artificial bank, though ever so old, can be here instantly detected. The Keltic and West-Saxon barrows still remain. The sites of the dwellings of the Britons are still plainly visible. The Roman potteries are untouched, and their urns, though lying but a few inches under the ground, unbroken. We can only very fairly conclude that, had there been houses, or villages, or churches destroyed, all trace of them would not be gone, nor entirely lost in the preserving record of local names.

It has, I am aware, been urged that since the Old-English churches were chiefly built of wood, we are not likely to find any ruins. This may be so. But by no process of reasoning can the absence of a thing prove its former presence. Nor need we pay much attention to the argument drawn from such names as Castle Malwood, The Castle near Burley, Castle Hill on the banks of the Avon, Lucas Castle, and Broomy and Thompson Castles in Ashley Walk. If these names prove anything, it is that there were a vast number of castles in the Forest, and very few churches. But Castle Malwood was standing long after the afforestation; whilst the Castle at Burley, and Castle Hill, and the others, were merely earthen fortifications and entrenchments, made by the Kelts and West-Saxons. Nor must we be led away by the few Forest names ending in ton, the Old-English tun, which, after all, means more often only a few scattered homesteads than even a village, still less a town or city, in the modern sense of the word.[24]

If, however, we look at the district from another point of view, we shall find further evidence against the Chroniclers. It was a part of the Natan Leaga[25]—a name still preserved in the various Netleys, Nateleys, and Nutleys, which remain—the Ytene of the British, that is, the furzy district, a title eminently characteristic of the soil.[26] Again, too, the villages and manors, such as Lyndhurst, Brockenhurst, Ashurst, and half a dozen more hursts, point to the woody nature of the place. Such names, also, as Roydon, the rough ground; Bramshaw, the bramble wood; Denny, the furzy ground; Wootton, the Odetune of *Domesday*; Stockeyford and Stockleigh, the woody place; Ashley, the ash ground; besides Staneswood, Arnwood, and Testwood, all more or less afforested in *Domesday*, clearly show its character.

After all, the best evidence is not from such arguments, but in the simple fact that the New Forest remains still the New Forest. Had the land been in any way profitable, modern skill, and capital, and enterprise, would have certainly been attracted. But its charms lie not, and never did, in the richness of its soil, but in its deep woods and wild moors.[27]

Our view of the matter, then, is that William, like all Normans, loving the chase—loving, too, the red deer, as the Old-English

Chronicler, with a sneer, remarks, as if he was their own father—converted what was before a half-wooded tract, a great part of which was his own as demesne, and the whole as prerogative, into a Royal Forest, giving it the name of the New Forest, in contradistinction to its former title of Ytene. To have laid waste a highly-cultivated district for the purposes of the chase, as the Chroniclers wish us to believe, would have defeated his chief object, as there would have been no shelter then, nor for many years to come, for the deer: and is contradicted, as we have seen, both by *Domesday*, by the very nature of the soil, and the names of the places.

The real truth is, that the stories, which fill our histories, of William devasting the country, burning the houses, murdering the people, have arisen from a totally wrong conception of an ancient forest. Until this confusion of an old forest with our modern ideas is removed, we can have no clear notions on the subject. Without at all leaning on the etymology which has been before given,[28] we must remember that an ancient forest did not simply mean a space thickly covered with trees, but also wild open ground, and lawns and glades.[29] The word hurst, which, as we have seen, is so common a termination throughout the district, means a wood which produces fodder for cattle, answering to the Old High-German *spreidach*.[30] The old forests possessed, if not a large, some scattered population. For them a special code of laws was made, or rather gradually developed itself. Canute himself appointed various officers—Primarii, our Verderers; Lespegend, our Regarders; and Tinemen, our Keepers. The offences of hunting, wounding, or killing a deer, striking a verderer or regarder, cutting vert, are all minutely specified in his Forest Law, and punished, according to rank and other circumstances, with different degrees of severity.[31] The Court of Swanimote was, in a sense, counterpart to the Courts of Folkemote and Portemote in towns. A forest was, in fact, a kingdom within a kingdom, with certain, well-defined laws, suited to its requirements, and differing from the common law of the land. The inhabitants had regular occupations, enjoyed, too, rights of pasturing cattle, feeding swine, and cutting timber.[32] All this, as we have seen, went on as before, not so much, but still the same, in the New Forest. Manors, too, with the exception of being subject to Forest Law, remained in the heart of it unmolested. According to the Chroniclers themselves, some rustics living on the spot convey, with a horse and cart, the bleeding body of Rufus to Winchester. According to them[33] also the King, previous to his death, feasted, with his retinue of servants, and huntsmen, and priests, and guests, at Castle Malwood, implying some means in the neighbourhood to furnish, if not the luxuries, the necessities of life. In *Domesday* we find, too, a keeper of the king's house holding the mill at Efford; also implying, at least, in a very different part of the Forest, a neighbourhood which could not have been quite destitute and deserted.[34] At a later period, when the Forest Laws had reached their climax of oppression, persons in the Forest, as we learn from Blount and the Testa de Nevill, hold their lands at Brockenhurst and Eyeworth,[35] by finding provisions for the king and fodder for his horse. But more than all, *Domesday*, corroborated as it is by the physical peculiarities of the country, by the evidence, too, of local names, by the Norman doorways, and pillars and arches at Fawley, and Brockenhurst and Milford, proves most distinctly—and most distinctly because so circumstantially—that the district was neither devastated, nor the houses burnt, nor the churches destroyed, nor the people murdered.

Some wrong, though, was doubtless committed: some hardships undergone. Lands, however useless, cannot be afforested without the feelings of the neighbourhood being outraged. And the story, gathering strength in proportion as the Conqueror and his son William the Red were hated by the conquered, at last assumed the tragical form which the Chroniclers have handed down to us, and modern historians repeated.

William's cruelty, however, lay not certainly in afforesting the district: it consisted rather in the systematic way in which he strove to reduce the English into abject slavery; in the fresh tortures with which he loaded the Danish Forest Laws; and in making it far better to kill a man than a deer. For these exactions was it that his family paid the penalty of their lives; and the retribution befel them there, where the superstitious West-Saxon would, above all others, have marked out as the spot fitted for their deaths.

View in Gibb's Hill Wood.

Footnotes

1. *England under the Anglo-Norman Kings.* Ed. Thorpe, p. 214.
2. The same, p. 266.
3. *The Chronicle.* Ed. Thorpe. Vol. i. p. 354. This, of course, must not be too literally taken. It is one of those stock phrases which so often recur in literature, and may be found, under rather different forms, applied to other princes.
4. Voltaire was the first to throw any doubt on the generally received account (*Essai sur les Moeurs et l'Esprit det Nations*, tom.

iii. ch. xlii. p. 169. *Pantheon Litteraire*. Paris, 1836). He has in England been followed by Warner (*Topographical Remarks on the South-Western Parts of Hampshire*, vol. i. pp. 164-197), and Lewis, in his *Historical Enquiries concerning the New Forest*, pp. 42-55.
5. Concerning the King's prerogative to make a forest wherever he pleased, and the ancient legal maxim that all beasts of the chase were exclusively his and his alone, see Manwood—*A Treatise of the Lawes of the Forest*, ch. ii. ff. 23-33, and ch. iii. sect. i. f. 33, 1615. We must remember, too, that, before the afforestation, as we learn from *Domesday*, William not only owned as his demesne, inherited from Edward the Confessor, a great deal of land in the district—at Ashley, Bashley, Hubborn, Wootton, Pilley, &c, besides nearly the whole of the Hundred of Boldre, but kept some—as at Eling, Breamore, and Ringwood—in his own hands.
6. Bouquet. *Recueil des Historiens des Gaules et de la France*, tom. xi., pref, No. xii. p. 14; and tom. xii., pref., No. xlix. pp. 46-48. Some account of him may be found in tom. x. p. 184, foot-note a, and in the preface of the same volume, No. xv. p. 28. See also preface to tom, viii., No. xxxi., p. 24, as also p. 254, foot-note a.
7. *De Ducibus Normannis*, book vii. c. ix.; in Camden's *Anglica Scripta*, p. 674.
8. *Chronicon ex Chronicis*. Ed. Thorpe. Vol. ii. p. 45. Published by the English Historical Society.
9. *Historia Ecclesiastica*, pars, iii., lib. x., in the *Patrologiæ Cursus Completus*. Ed. J. P. Migne. Vol. clxxxviii. p. 749 c. Paris, 1855.
10. *De Nugis Curialium Distinctiones Quinque*, distinc. v. cap. vi. p. 222. Published by the Camden Society.
11. *De Eventibus Angliæ*, lib. ii. cap. vii., in Twysden's *Historia Anglicanæ Scriptores Decem*, p. 2373. I am almost ashamed to quote Knyghton, but it is as well to give the most unfavourable account. Spotswood, in his *History of the Church of Scotland* (book ii. p. 30, fourth edition, 1577), repeats the same blunder as Walter Mapes and Knyghton, adding that the New Forest was at

Winchester, and that Rufus destroyed thirty churches.
12. For the sake of brevity, let me add that William of Malmesbury (*Gesta Rerum Anglorum*, vol. ii. p. 455, published by the English Historical Society, 1840), Henry of Huntingdon (*Historiarum*, lib. vi., in Savile's *Rerum Anglicarum Scriptores*, p. 371), Simon of Durham (*De Gestis Regum Anglorum*, in the *Historiæ Anglicanæ Scriptores Decem*, p. 225), copying word for word from Florence, Roger Hoveden (*Annalium Pars Prior, Willielmus Junior*, in the *Rerum Anglicarum Scriptores*, p. 468), Roger of Wendover (*Flores Historiarum*, vol. ii. pp. 25, 26, published by the English Historical Society), Walter Hemingburgh (*De Gestis Regum Angliæ*, vol. i. p. 33, published by the English Historical Society), and John Ross (*Historia Regum Angliæ*, pp. 112, 113. Ed. Hearne. Oxford, 1716), repeat, according to their different degrees of accuracy, the general story of the Conqueror destroying villages and exterminating the inhabitants.
13. *The Chronicle*. Ed. Thorpe, as before quoted. Nor does the writer, when another opportunity presents itself at Rufus's death, mention the matter, but passes it over in significant silence. The same volume, p. 364.
14. See *Domesday* (the photo-zincographed fac-simile of the part relating to Hampshire; published at the Ordnance Survey Office, 1861), p. xxix. b, under Bertramelei, Fistelslai, Odetune, and Oxelei.
15. See in *Domesday*, as before, p. xxvii. b, the entry under Langelei—"Aluric Petit tenet unam virgatam in Forestâ." See, too, p. iii. b, under Edlingas.
16. See in *Domesday*, under Thuinam, Holeest, Slacham, Rinwede, p. iv. a; and Herdel, p. xxviii. b.
17. See in *Domesday*, out of many instances, Esselei and Suei, p. xxix. b; Bailocheslei, p. xiv. b; Wolnetune and Bedeslei, p. xxviii. a; Hentune, p. xxviii. b; and Linhest, p. iv. a.
18. It is possible that whilst the survey was being taken Saulf

died. If this be so, we find an instance of feeling in allowing his widow to still rent the lands at Hubborn, which could little have been expected. The name seems to have been misspelt in various entries. See *Domesday*, p. xxix. b, under Sanhest and Melleford.
19. Aluric is probably the physician of that name mentioned in *Domesday*, p. xxix. a, as holding land in the hundred of Egheiete. Not to take up further space, let me here only notice some of the Old-English names of persons in *Domesday* holding lands in places which had been more or less afforested, such as Godric (probably Godric Malf) at Wootton, Willac in the hundred of Egheiete, Uluric at Godshill, in the actual Forest, and Wislac at Oxley. See *Domesday* under the words Odetune, Godesmanescamp, and Oxelei, p. xxix. b. See, also, under Totintone, p. xxvii. a, where Agemund and Alric hold lands which the former, and the latter's father, had held of Edward.
20. Passing over the later and more highly-coloured accounts, we will content ourselves with Florence of Worcester, as more trustworthy, whose words are—"Antiquis enim temporibus, Edwardi scilicet Regis, et aliorum Angliæ Regum predecessorum ejus, hæc regio incolis Dei et ecclesiis nitebat uberrime." (Thorpe's edition, as before quoted.) Were this, even in a limited degree, true, the Forest would present the strange anomaly of possessing more churches then than it does now, with a great increase of population. The *Domesday* census, we may add, makes the inhabitants of that portion which is called "In Novâ Forestâ et circa eam," a little over two hundred. See Ellis's *Introduction to Domesday*, vol. ii. p. 450.
21. In support of these statements, I may quote from the Prize Essay on the Farming of Hampshire, published in the *Journal of the Royal Agricultural Society of England* (vol. xxii., part ii., No. 48, 1861), and which was certainly not written with any view to historical evidence, but simply from an agricultural point. At pp. 242, 243, the author says: "The outlying New Forest block consists

of more recent and unprofitable deposits. This tract appears to the ordinary observer, at first sight, to be a mixed mass of clays, marls, sands, and gravels. The apparent confusion arises from the variety of the strata, from the confined space in which they are deposited, and from the manner in which, on the numerous hills and knolls, they overlie one another, or are concealed by drift gravel." And again, at pp. 250, 251, he continues: "Of the Burley Walk, the part to the west of Burley Beacon, and round it, is nothing but sand or clay, growing rushes, with here and there some 'bed furze.' . . . . The Upper Bagshots, about Burley Beacon, round by Rhinefield and Denney Lodges, and so on towards Fawley, are hungry sands devoid of staple:" and finally sums up by saying, "half of the 63,000 acres are not worth 1*s*. 6*d*. an acre," p. 330.
22. In that portion under "In Novâ Forestâ et circa eam."
23. Warner, vol. ii. p. 33, says Hordle Church was standing when *Domesday* was made. This is a mistake. It was, however, built soon after, as we know from some grants of Baldwin de Redvers.
24. Mr. Thorpe notices, in his edition of *The Chronicle*, vol. ii. p. 94, footnote, its early use, in a document of Eadger's, a.d. 964, in the sense of a town; but in the first place it certainly meant only an inclosed spot. There appears to have been at some time, in the south part of the Forest, a church near Wootton, the Odetune of *Domesday*, where its memory is still preserved in the name of Church Lytton given to a small plot of ground. Rose, in his notes to the *Red King*, suggests that Church Moor and Church Place indicate other places of worship. Church Moor is a very unlikely situation, being a large and deep morass, and could well, from its situation, have been nothing else, and, in all probability, takes its name, in quite modern times, from some person. But Church Place at Sloden, like Church Green in Eyeworth Wood, is certainly merely the embankments near which the Romano-British population employed in the Roman potteries, once lived, and which ignorance and superstition have turned into sacred

ground. The word Lytton, at Wootton, however, makes the former position certain, but by no means necessitates that the church was standing at the afforestation. Thus we know that in Leland's time a chapel was in existence at Fritham (*Itinerary*, ed. Hearne, vol. vi. f. 100, p. 88), which has since his day disappeared. It would, of course, be absurd to argue that all ruins which have been, or yet may be found, were caused by the Conqueror. Rose's *Red King* was privately printed, and I know the book only through Ellis's *Introduction to Domesday* (vol. i. p. 108), and a notice of it in the *Edinburgh Review* (Jan., 1809, vol. xiii. pp. 425, 426); but it is amusing to see certain recent writers trying to prove William's devastion because the remains of brick have been discovered. This certainly shows that long since the Conqueror's time the people have endeavoured, with very ill-success, to live on the barren soil of the Forest. I may, perhaps, add that Mr. Ackerman, the well-known archaeologist, when, a few years since, exploring the Roman potteries in the Forest (for which see chapter xvii.), in vain tried there, or in other parts, to find any traces of old buildings. (*Archaeologia*, vol. xxxv. p. 97.)

25. See Dr. Guest's *Early English Settlements in South Britain*; *Proceedings of the Archaeological Institute*, Salisbury volume, p. 57.

26. "Nova Foresta, quæ linguâ Anglorum Ytene nuncupatur," however, says Florence of Worcester (vol. ii. pp. 44, 45, ed. Thorpe); but the Keltic origin of the word is better.

27. The names of the fields in the various farms adjoining the Forest—Furzy Close, Heathy Close, Cold Croft, Starvesall, Hungry Hill, Rough Pastures, &c. &c.—are not without meaning. The common Forest proverb of "lark's-lees," applied to the soil, pretty clearly, too, shows its quality.

28. See chapter ii. p. 10, footnote. There are a number of derivations given for the word, but none are satisfactory.

29. Manwood defines a forest "a certaine territorie of woody grounds and fruitful pastures." *A Treatise of the Lawes of the*

Forest. London, 1619. Chap. i. f. 18.
30. See Mr. Davies's paper on the Races of Lancashire, *Transactions of the Philological Society,* 1855, p. 258. In *Domesday,* as before, under Clatinges, p. xviii. a, we find, "Silva inutilis," that is, a wood, I should suppose, which has no beech, oak, ash, nor holly, but only yews or thorns. Again, under Borgate, p. iv. b, we find, "Pastura quæ reddebat xl. porcos est in forestâ Regis." The woods, as before mentioned, at pp. 11, 12, foot-note, are always, in *Domesday,* rated by the number of swine they maintain.
31. See Manwood, as before, ff. 1-5.
32. In the Charta de Forestâ of Canute (Manwood, f. 3, sect. 27) mention is made in the forests of horses, cows, and wild goats which are all protected; and from sect. 28 it is plain that, under certain limitations, people might cut fuel. These, with other privileges, such as killing game on their own lands (see sect. xxx. f. 4)—for, by theory, all game was the King's—were compensations given to the forester for being subject to Forest Law.
Further, from the Charta de Forestâ of Henry III. (Manwood, ff. 6-11), we find that persons had houses and farms, and even woods, in the very centre of the King's forests; and the charter provides that they may there, on their own lands, build mills on the forest streams, sink wells, and dig marl-pits, referring, most probably, in the last case, to the New Forest, where marl has been used, from time immemorial, to manure the land; and, further, that in their own woods, even though in the forest, they might keep hawks, and go hawking. (See f. 7, sects, xii., xiii.)
It shows, too, that there was a population who gained their livelihood, as to this day, by huckstering, buying and selling small quantities of timber, making brushes, and dealing in bark and coal, which last article evidently points to the Forest of Dean. (F. 7, sect, xiv.)
We must not imagine that the Charta de Forestâ of Henry III. was entirely a series of new privileges. They were, with some notable

exceptions, simply those rights which had been received from the earliest times in compensation for some of the hardships of the Forest Laws, and which had been wrested away, probably by Richard or John, but which had never been granted to those who dwelt outside the Forest. (On this point see especially "Ordinatio Foreste," 33rd Edward I., *Statutes of the Realm*, vol. i. p. 144. And again, "Ordinatio Foreste," 34th Edward I., sect, vi., same volume, p. 149, where the rights of pasturage are re-allowed to those who have lost it by the recent perambulation made in the twenty-ninth year of the King's reign.)

I think we may, therefore, gain from these clauses, especially when taken in conjunction with those of the Charta de Forestâ of Canute, a tolerably correct picture of an ancient forest—that it consisted not merely of large timber and thick underwood, a cover for deer, but of extensive plains,—still here preserved in the various *leys*—grazed over by cattle, with here and there cultivated spots, and homesteads inhabited by a poor, but industrious, population.

33. See chapter ix.

34. See *Domesday*, as before, p. xxix. b., under Einforde.

35. See chapters vii. and x.

CHAPTER IV.

ITS LATER HISTORY.

The Millaford Brook, Haliday's Hill Wood

We need not dwell so long upon this as the former portion of the History, for in many cases it is nothing but a bare recital of perambulations and Acts of Parliament. The true history of a forest is rather an account of its trees and its flowers and birds, than an historical narrative. Yet even here there are some important facts connected with the nation's life, and illustrating the character of its kings.

We meet with no perambulation of the New Forest until the eighth year of Edward I.—the second ever made of an English forest—and, by comparing it with *Domesday*, we may see how, since the Conqueror's time, the Forest had gradually taken the natural limits of the country—the Avon and the Southampton Water bounding it on the east and west, and the sea on the south, and the chalk of Wiltshire on the north.[1]

The next perambulation in the twenty-ninth year of the same realm is more noticeable,[2] as it disafforests so much. It is the same perambulation which we find made in the twenty-second year of Charles II., and nominally the same which is followed to this day.

To understand the cause of the difference in these perambulations, we must, in fact, thoroughly understand the great movements which had been going on during the previous years, and the increasing power of the nobles and the people. From Henry III. had been wrung the Charta de Forestâ, the terms of which had been settled before John's death. Still, little, or scarcely anything, was put into practical effect. In 1297, however, the Earls of Hereford and Norfolk not only refused to accompany Edward I. to Flanders, but, upon their suspension from their offices, issued a proclamation, complaining that the two Charters of the liberties of the people were not observed. On the 10th of October, a Parliament was assembled, and his son passed the "Confirmatio Cartarum," to which Edward, now at Ghent, assented. Still the two earls, from various causes, were not satisfied; and in 1298 demanded that the perambulations of the different Forests should be made. In consequence, during the summer of the next year, the King issued writs to the sheriffs, promising that the commissioners should meet about

Michaelmas at Northampton.[3]

This was done: and the perambulation of the New Forest was carried out in strict accordance with the provisions of the Charta de Forestâ, for the jurors who were employed expressly state that the bounds which they have determined were those of the Forest before the reign of Henry II.; and that all those places mentioned in the perambulation of 1279 and now omitted, were afforested by his successors, though they cannot say to what extent or by whom.[4] Most probably it had been reserved for John to show here, as in other cases, to what absolute madness selfishness will carry a man.

After this, nothing, with one exception, of any general importance occurs.[5] Having in his prosperity incurred all the odium of attempting to revive the hated Forest Laws, in his adversity Charles I. granted as security the New Forest, and Sherwood, and other Crown lands to his creditors.[6] He had still learnt no lesson from the Ship-money, and would have pawned England itself, rather than yield to that obstinacy, which was but the other side of his weakness of character.

With the decline of hawking and hunting, the Forest Laws fell into decay, and the Forests themselves were less regarded, and their boundaries less strictly observed. Under the Stuarts, we find the first traces of that system, which at last resulted in the almost entire devastation of the New Forest. James I. granted no less than twenty assart lands—agri exsariti—there having been previously only three;[7] and gave the privilege of windfalls to various persons;[8] whilst officers actually applied to him for trees in lieu of pay for their troops:[9] and Charles II. bestowed the young woods of Brockenhurst to the maids of honour of his court.[10]

Manwood, who wrote towards the end of Elizabeth's reign, had, long before this, predicted what must happen, and the straits to which the English navy, as we know was the case, would be reduced. In Charles I.'s time the Forests were in a shameful condition. The keepers were in arrears of wages, and paid themselves out of the timber.[11] The consequences soon came. There was nothing left but wind-shaken and decayed trees in the New Forest, quite unfit for building ships,[12] Charles II., however, in 1669, probably influenced by Evelyn's *Silva*, which appeared four years before, and had given a great impulse, throughout England, to planting, enclosed three hundred acres as a nursery for young oaks. But the waste and devastation still continued. At last, William III. legislated on the subject, for, to use the words of the Act, "the Forest was in danger of being destroyed;"[13] and power was given to plant six thousand acres. In 1703 came the great hurricane, which Evelyn so deplores, uprooting some four thousand of the best oaks.

Nothing was done towards planting during the reigns of Anne and George I.;[14] and Phillipson's and Pitt's plantations in 1755 and 1756 are the next, but they have never thrived, owing to the land not having been drained, and the trees not having been thinned out at the proper time.

In 1789 a Commission was appointed, and revealed a terrible state of things. William's provisions had not only been set aside, but defied. Cattle were turned out, the furze and heath cut, and the marl dug by those who had no privileges. The Forest was, in fact, robbed under every pretext. The deer, from being overstocked, died in the winter by hundreds from starvation. On every side, too, encroachments were made by those whose business it was to prevent them. The rabbits destroyed the young timber, whilst the old was stolen.[15]

In 1800 there was fresh legislation,[16] but it does not seem to have taken much effect; though, in 1808, a new system of planting upon a definite plan was introduced.

In 1848 another Commission was appointed, and showed that the old abuses still lingered, that depredations were still committed, and encroachments still made.[17] Law was at last restored. A great number of the claims were disallowed, and the rights of the commoners defined. So many head of cattle may now be turned out, by those who have Forest rights, through the year, except during the fence-month, which lasts from the 20th of June to the 20th of July, and the winter-hayning, from the 22nd of November to the 4th of May. Pigs, too, upon a nominal payment, may also be turned out for the mast and acorns during the pannage-month, lasting from September the 25th to the 22nd of November, by those who have a right to common of pannage; whilst any person can, by applying to the woodmen, buy wood for fuel.

Lastly, in 1851, the deer, the cause of so much ill-feeling and crime, were abolished, and the Crown thereby acquired the right of planting 10,000 additional acres.[18] These changes have already effected much good both for the district and the inhabitants. The enclosures are now systematically drained; and the Foresters find, in the works which are being carried forward, regular employment throughout the year.[19] A large nursery has been formed at Rhinefield, and somewhere about 700 acres are annually planted, the young oaks being set between Scotch firs, which serve both as "nurses" to draw them up, and a screen to shelter them from the winds. Experiments, too, are being made to acclimatize several new trees, but it is premature to judge with what success.

Further, I need scarcely add that all sorts of schemes, from the day when Defoe proposed to colonize the district with the Palatine refugees from the Rhine to the present, have been suggested for reclaiming the Forest. None have ever, from the nature of the soil, been found to answer; and the present condition is certainly, for many reasons, the best. The time will some day arrive when, as England becomes more and more overcrowded,—as each heath and common are swallowed up,—the New Forest will be as much a necessity to the country as the parks are now to London. We talk about the duty of reclaiming waste lands, and making corn spring up where none before grew. But it is often as much a duty to leave them alone. Land has higher and nobler offices to perform than to support houses or grow corn—to nourish not so much the body as the mind of man, to gladden the eye with its loveliness, and to brace his soul with that strength which is alone to be gained in the solitude of the moors and the woods.

The Woodcutter's Track, between Mark Ash and Winding Shoot.

Footnotes

1. The following translation is made from the original in the Record Office. Southt Plita Foreste, A° viii.° E. I.mi "The metes and boundaries of the New Forest from the first time it was

afforested. First, from Hudeburwe to Folkewell; thence to the Redechowe; thence to the Bredewelle; thence to Brodenok; thence to the Chertihowe; thence to the Brygge; thence to Burnford; thence to Kademannesforde; thence to Selney Water; thence to Orebrugge; thence to the Wade as the water runs; thence to the Eldeburwe; thence to Meche; thence to Redebrugge as the bank of the Terste runs; thence to Kalkesore as the sea runs; thence to the Hurste, along the sea-shore; thence to Christ Church Bridge as the sea flows; thence as the Avene extends, as far as the bridge of Forthingebrugge; thence as the Avene flows to Moletone; thence as the Avene flows to Northchardeford and Sechemle; and so in length by a ditch, which stretches to Herdeberwe." It is this old natural boundary which, as stated in the preface, we have adopted for the limits of the book. A copy of the original may be found in the *Journals of the House of Commons*, vol. xliv., appendix, p. 574, 1789.

2. This may also be found, with the perambulation made in the twenty-second year of Charles II., in the *Journal of the House of Commons*, vol. xliv., appendix, pp. 574, 575, 1789. It is also given in Lewis's *Historical Enquiries upon the New Forest*, appendix ii. pp. 174-177.

3. This is not the place to say more on this most important chapter of English history. See, however, on the subject, *The Great Charter: and the Charter of the Forest*, by Blackstone, Introduction, pp. lx.-lxxii. 1759. For the oppressions which still existed under the shelter of the Forest Laws, see the preamble to the "Ordinatio Foreste," 34th Edward I., *Statutes of the Realm*, vol. i. p. 147.

4. "Quid et quantum temporibus cujuslibet regis nullo modo eis constare potest." The conclusion of the perambulation. Some little difficulty attends these perambulations. From *Domesday*, it is certain that the Conqueror afforested land on the west of the Avon at Holdenhurst, Breamore, and Harbridge. And amongst the MSS. of Lincoln's Inn Library we find a copy of a charter of William

of Scotland, dated, curiously enough, "Hindhop Burnemuth, in meâ Novâ Forestâ, 10 Kal. Junii, 1171." (See Hunter's "*Three Catalogues*" &c, p. 278, No. 78, 1838.) It would seem, from what Edward's commissioners say, that these afforestations, which had taken place since Henry II.'s time, were all made inside the actual boundaries of the Forest. It has been generally supposed that the perambulation in the eighth year of Edward I. was the first ever made of an English forest. This is not the case; for in the Record Office, in the Plìta Foreste de Cõm. Southt. LIIItio R. H. III., No. III., may be found the perambulation of a forest in the north of Hampshire.

5. For a good account of all details connected with the history of the New Forest, see the Sub-Report by the Secretary of the Royal New and Waltham Forest Commission, *Reports from Commissioners* (11), vol. xxx. pp. 267-309, 1850; and also the Fifth Report of the Land Revenue Commissioners in 1789, published July 24th of that year, to be found also in the *Journals of the House of Commons*, vol. xliv. pp. 552-571.

6. See "The humble petition of Richard Spencer, Esq., Sir Gervas Clifton, Knight and Baronet, and others, to enter upon the New Forest and Sherwood Forest," &c. &c. Record Office, Domestic Series, Charles II, No. 8. f. 26, July 21st, 1660.

7. MSS. prepared by Mr. Record-Keeper Fearnside, quoted in the Secretary's Sub-Report of the Royal New and Waltham Forest Commission, *Reports from Commissioners* (11), vol. xxx. p. 342.

8. See Grant Book at the Record Office, 1613, vol. 141, p. 127—"4th October, a Grant to Richard Kilborne, alias Hunt, and Thomas Tilsby (of) the benefitt of all Morefalls within the New Forest, for the terme of one and twenty years."

9. See "The humble petition of Captayne "Walter Neale" for "two thousand decayed trees out of the New Forest, in consideracion" of 460l., which he had advanced to his company engaged in Count Mansfeldt's expedition. Record Office. Domestic Series, No. 184,

Feb., 1625, f. 62.
10. See warrant from Charles II. to the Lord Treasurer Southampton, that "Winefred Wells may take and receive for her own use" King's Coppice at Fawley, and New Coppice and Iron's Hill Coppice at Brockenhurst. Record Office. Domestic Series, No. 96, April 1st, 1664, f. 16. Three years before this there had been a petition from a Frances Wells "to bestowe upon her and her children for twenty-one yeares the Moorefall trees in three walks in the New Forest . . . . and seven or eight acres of ground, and ten or twelve timber trees, to build a habitation." The petition was referred to Southampton, who wrote on the margin, "I conceive this an unfit way to gratify this petitioner, for under pretence of such Moorefall trees much waste is often committed." Record Office. Domestic Series, No. 34, April 2nd, 1661, f. 14. Hence the reason of Charles's warrant in the case of Winefred Wells, as he knew that the Lord Treasurer was so strongly opposed to any such grants.
11. See the report of Peter Pett, one of the King's master shipwrights, "Touching the fforests of Shottover and Stowood." Record Office. Domestic Series. No. 216, f. 56. i. May 10th, 1632. The New Forest, however, seems from this report to have been much better in this respect.
12. See "Necessarie Remembrances concerning the preservation of timber, &c." Record Office. Domestic Series. Charles I., No. 229, f. 114. Without date, but some time in 1632.
13. 9th and 10th of William III., chap, xxxvi., 1693. An abstract of the Act may be found in the *Journals of the House of Commons*, vol. xliv., appendix, pp. 576-578.
14. To show how for years the Forest was neglected and robbed, we find, from a survey made in James I.'s reign, 1608, that there were no less than 123,927 growing trees fit for felling, and decaying trees which would yield 118,000 loads of timber; whilst in Queen Anne's reign, in 1707, only 12,476 are reported as serviceable.

See Fifth Report of the Land Revenue Commissioners, *Journals of the House of Commons*, vol. xliv. p. 563. The waste in James I. and Charles I.'s time must have been enormous, for from the "Necessarie Remembrances" before quoted we find that there were not in 1632 much above 2,000 serviceable trees in the whole Forest.

15. See, as before, Fifth Report of the Land Revenue Commissioners, pp. 561, 562, and especially the evidence of the under-steward, Appendix, 583. As far back as February 20th, 1619, we find that James I. gave the Earl of Southampton 1,200l. a year as compensation for the damage which the enormous quantity of deer in the Forest caused to his land. Letter from Gerrard to Carleton, Feb. 20, 1618/1619, Record Office. Domestic Series, No. 105, f. 120. Gilpin (vol. ii. pp. 32, 33, third edition) states that in his day two keepers alone robbed the Forest to the value of 50,000l.

16. *Journals of the House of Commons*, vol. xlvii. pp. 611-792; vol. lv. pp. 600-784.

17. See the evidence in the *Parliamentary Papers*, 1849, Nos. 513, 538. Of the Forest Rights and Privileges, the secretary to the New Forest Commission writes: "The present state of the New Forest in this respect is little less than absolute anarchy." (*Reports of Commissioners* (11), vol. xxx. p. 357, 1850.) It should be distinctly understood, as was shown in the last chapter, that these Rights had their origin as a compensation to those whose lands had been afforested by the King, and who were, in consequence, subject to the Forest Laws, and the injury done by the deer. Now that the injury is no longer sustained, and the exercise of the Prerogative has ceased, so ought also the privileges. The Crown, however, has not pressed this, and the Rights are thus still enjoyed. *A Register of Decisions on Claims to Forest Rights*, with each person's name, and the amount of his privileges, was published in 1858.

18. The present statistics of the Forest are—Freehold estates, being private property, within the Forest boundaries, 27,140

acres; copyhold, belonging to her Majesty's manor of Lyndhurst, 125; leasehold, under the Crown, 600; enclosures belonging to the lodges, 500; freeholds of the Crown, planted, 1,000; woods and wastes of the Forest, 63,000: total, 92,365 acres. The value of timber supplied to the navy during the last ten years has been, on the average, nearly 7,000l. a year. The receipts for the year ending 31st of March, 1860, derived from the sale of timber, bark, fagots, marl, and gravel, and rent of farms and cottages, &c., were 23,125l. 6s. 6d.; whilst the expenses for labour, trees, carriage of timber, and salaries, were 12,913l. 1s. 7d; thus showing a considerable profit. (From the Thirty-eighth Report of the Commissioners of her Majesty's Woods and Forests.) The management of the Forest is now in the hands of a deputy-surveyor, three assistants, and eight keepers; whilst four verderers try all cases of stealing timber, turf, and furze.

19. See further, on the condition of the Forest population, chapters xv. and xvi. When stripping bark and felling timber in the spring, the men can earn considerably more than at other times. The average wages are two shillings a day for ordinary labourers, but all work, which can be, is done by the piece.

CHAPTER V.

CALSHOT CASTLE AND THE OLD SOUTH-EASTERN COAST.

Calshot Castle

THIS corner of the Forest, once perhaps the most beautiful, is now the least known, because, to most people, so inaccessible. It lies quite by itself. No railway yet disfigures its fields and dells. The best way to see it and the whole Forest is to cross the ferry at Southampton, and land at the hard at Hythe. And as we cross, behind us, amongst a clump of trees, rises the ruined west end of Netley Abbey Church, and the modern tower on Henry VIII.'s

Fort; whilst, lower down, the new Government Hospital loads the shore with all its costly ugliness. If we have not, perhaps, yet reached the height of Continental profanity which has turned the Convent of Cordova into barracks, and St. Bernard's Monastery at Clairvaux into a prison, and the Church of Cluny into racing stables, we yet seem to delight to place side by side with the noblest conceptions that ever rose in beauty from English ground, our modern abortions. There is not a cathedral town whose minster-square is not disgraced by some pretentious shed. And now Government not only invades the country, but chooses above all, the better to display our folly, that place which the old Cistercian monks had for ever made sacred by the loveliness of faith and work.

Hythe is only a little village, but as its name shows, once the port of the New Forest.[1] The Forest, however, has now receded from it, and in this chapter we shall see nothing of its woods. The district, however, is too important in an historical point of view to be omitted. The walk, even though it is not over wild moors and commons, is still very beautiful. True English lanes will lead us by quiet dells, with glimpses here and there through hedgerow elms of the blue Southampton water, down to the shore of the Solent.

So, leaving Hythe[2] and going southward, skirting Cadlands Park, we reach Fawley, the Falalie and Falegia of *Domesday*, where, at the time of the survey, Walchelinus, Bishop of Winchester, held one hyde and three yardlands. The whole of the village was thrown into the Forest, but in its place now are ploughed fields and grass pastures. The church, with its central tower, stands at the entrance of the village, and its handsome Romanesque doorway shows plainly that the Conqueror did not destroy every place of worship. The building was partially restored in 1844, but the pillars on the north side of the chancel were copied from the original Norman work, which, with the three piscinas and the hagioscope, give it a further interest to the ecclesiologist.

After Fawley, the walk becomes more beautiful. We pass deep lanes and scattered cottages set in their trim gardens, when suddenly on the shore rises the round gray castle of Calshot, standing at the very end of a bar of sand, separating the Southampton water from the Solent. Though much repaired, it stands not much altered from Henry VIII.'s original blockhouse. Once of great importance, its garrison now consists of only the coastguard and a master-gunner. Its walls are still strong, measuring in the lower embrasures sixteen feet through, but the upper storeys are much slighter. On the west side is cut the date 1518, whilst some stone cannon balls of the Commonwealth period show the importance Cromwell attached to the place. But the stronger fortifications of Hurst, and the new batteries in the Isle of Wight, have done away with its necessity, and it stands now only as a monument of Tudor patriotism and of Cromwell's care.[3]

But the place has older associations than these. In *The Chronicle* and *Florence of Worcester* we read[4] that, in 495, Cerdic and his son Cynric arrived with five ships, and landed at Cerdices-ora, and on the same day defeated the natives. No site has given rise to so much discussion as this Cerdices-ora. Mr. Thorpe in one place says it is not known, whilst in another, by an evident oversight, he fixes upon Charford.[5] Dr. Guest places it at the mouth of the river Itchen,[6] whilst Mr. Pearson and others have identified it with Yarmouth.[7] Now, I think there can be little doubt, looking both at the etymology of the name and the situation, that Calshot is the true place. The land here runs out into the sea with no less than ten fathoms of water close to it, so that large vessels can to this day lie alongside the Castle. It is the first part, too, of the mainland which can be reached, and on its lee side offers a safe anchorage. Besides, about four miles off stand some barrows, which, though we may not be able to identify them as covering those slain in the first battle which the West-Saxons fought, offer some presumption in favour of that theory. In the very word Calshot, and its intermediate forms of Caushot, Caldshore, and Cauldshore,[8] we may, without difficulty, recognize a corruption of the original Cerdices-ora of the Chronicle and Florence. The word is formed like the names of various places close by, such as Needsore (the under-shore) and Stansore Point.[9] But going farther back, we come much nearer to its original form in the old Forest perambulation made in the eighth year of Edward I., where it is spelt Kalkesore.[10] As then, Charford, on the north-east borders of the New Forest, is the representative of Cerdices-ford, where Cerdic's last victory was gained over Ambrosius; so here, I think, at the south-west, near Kalkesore, now Calshot, was his first achieved.

From this point the scenery completely changes. Instead of lanes and cultivated fields, the shingly beach of the Solent, covered in places to the water's edge with woods, sweeps away to the west. Passing on to Eagle-hurst, and noticing the truth of the termination even to this day, let us sit down on the shore. Here is a view which should be remembered. In one sense the world cannot show its equal. Far away to the east stretches the low Hampshire coast, ended by the harbour of Portsmouth and its bare forest of masts. To the south, towards Spithead, rides the long line of battle-ships; and round the harbours of the two Cowes sail fleets of yachts, showing how much still of the old Scandinavian blood runs in our veins—of the spirit which finds pleasure in adventure and delight in danger. Steamers, with their black pennants of smoke, hurry down the narrow strait, carrying the news or the merchandise of the world; whilst all is overshadowed by supreme natural beauty, the hills of the Isle of Wight standing boldly up, crested with their soft green downs, and their dark purple shadows resting fold over fold on the valley sides. Still continuing along the shore we reach Leap, a small fishing village, where boats ply across from its hard to the Island. Its name is derived from the Old-English leap, a weel, or basket for catching fish. Here, it is said, but I know not on what authority save that worst—tradition, that the Dauphin, afterwards Louis VIII. of France, embarked after the defeat of his army at Lincoln, and his fleet off Dover. Certain it is that he had adherents to his cause in the neighbourhood, especially in William de Vernon, whose arms were formerly blazoned with his own in the east window of the north aisle of the Forest Church of Boldre.[11]

On somewhat better authority,[12] it rests that the unhappy Charles I., on the 18th of November, 1647, outwitted by his enemies and deceived by his friends, entrusted himself, after his flight from Hampton Court, to Colonel Hammond, and, embarking here, returned by Hurst to atone for the past by his life.

But of greater interest is the Roman Road which connected Leap with Southampton and Winchester in one direction, and Ringwood and the west in another. Its traces may be found not only here but on the opposite side, where, still known by the Norman name of Rue Street, it passes westward of Carisbrook to the extreme south of the Island. Its old appellation is preserved, too, on this side in the name of a farmhouse—King's Rue, and Rue Copse, and Rue Common; and it is well worthy of notice that this word is even now sometimes used in the Forest, as in Sussex, for a row or hedgerow. The road, however, can still tell us something of the past. The opinion of late philologists and geographers, with the exception of Lappenburg and Sir G. C. Lewis, has been against the idea that the Isle of Wight was the Vectis or Ictis of the ancients. The arguments, however, against the passage in Diodorus Siculus,[13] that it would be so much easier for the Phoenicians to have exported the tin from the Cassiterides instead of bringing it by inland transport to the island, and then shipping it to Gaul, is founded upon ignorance.

Sea carriage was then far more difficult and dangerous than land conveyance. Ancient mariners were easily frightened, and their vessels put into land every night. As Sir G. C. Lewis further remarks, foreign merchants were always regarded with jealousy and distrust, and the overland route would enable the traffic to be carried out through the whole distance by native traders.[14]

Singularly enough, however, Warner[15] states that a large mass of tin was found on the very site of this old Roman road. Not only, too, was tin brought here from Cornwall, but also lead from the Mendip Hills. Pigs of it have been picked up on a branch of the same Roman road running from Uphill on the Severn to Salisbury, and from thence joining the Leap road. One of them, stamped with the name of Hadrian, is now in the Bath Museum. We are thus enabled to connect Leap with the famous passage of the Greek historian.

Sir George Lewis's theory has, too, been singularly corroborated in other directions, especially by the large quantities of bronze ornaments found during the excavations in the Swiss Lakes, 1858 and 1854, the metals of which could only have been brought there by an overland route.

Further, too, we must not reject the account of Diodorus, because he says that at low tide the tin was carried over in carts. We must remember the extremely indefinite views of the ancients on all geographical subjects. The vaguest ideas were held, especially about Britain. Erring in a different direction, the mistake is not so bad as Pliny's, in making the Island six days' sail from England. There seems, however, a most natural explanation, that Diodorus, not having been there, took for granted the wild traditions and rumours which reached him, and which, even in these days, with only the slightest possible variation of form, still hold their ground with the Forest peasantry, in the legend that the stone of which Beaulieu Abbey is built was brought over the dry bed of the Solent, in carts, from the Binstead Quarries.

Still the passage is not without the further difficulty, that Diodorus seems, from the context, to have supposed that the Island was situated close to where the tin was dug. This, again, must be set down to that ignorance of geography, which has involved all Greek writers in such extraordinary mistakes.

Leap itself is now nothing but a village, with a scattered agricultural population; some few, however, maintaining themselves by fishing in the summer, and in the winter by shooting the ducks and geese which flock to the creeks and harbours of the Solent. Leaving it, and still keeping westward, we come to the Beaulieu river, where, in the autumn, after the heavy floods from the Forest, the salmon leap and sport in the freshets. The road now winds past Exbury by the side of thick copses which fringe the river. At last, at Hill Top, we reach Beaulieu Heath, and, in the far distance, the green foliage of the Forest hangs cloudlike in the air, whilst down in the valley lies the village of Beaulieu.

The Norman Doorway of Fawley Church.

Footnotes

1. In the *Rolls of Parliament*, vol. i. p. 125, A.D. 1293, 21st Edward I., is an account of a vessel, the All Saints, "de Hethe juxta Novam Forestam," which, laden with wine from Rochelle, was wrecked and plundered on the Cornish coast.
2. A little beyond Hythe is a good example of Mr. Kemble's test (see the *Saxons in England*, vol. i., Appendix A, p. 481) for recognizing the Ancient Mark. To the north lies Eling, the Mark of the Ealingas, and in regular succession from it come the various hursts, holts, and dens, now to be seen in Ashurst, Buckholt, and Dibden. The last village has a very picturesque church, its roof

completely thatched with ivy, disfigured, however, by a wretched spire. A Roman glass manufactory has, I believe, whilst these pages were in the press, been here discovered. In *Domesday* it possessed a saltern and a fishery. Eling, at the same time, maintained two mills, which paid twenty-five shillings, a fishery and a saltern, both free from tax. The manor was bound, in the time of Edward the Confessor, to find half-a-day's entertainment (*fima*) for the King. For a curious extract from its parish register, see chapter xix. Staneswood (Staneude), which is more southward, also, according to *Domesday*, possessed a mill which paid five shillings, and two fisheries worth fifty pence. Farther north lies Redbridge, the Rodbrige of *Domesday*, which also maintained two mills, assessed, however, at fifty shillings. This was the Hrcutford and Vadum Arandinis of Bede, where lived Cynibert the Abbot, who, failing in his attempt to save the two sons of Arvald from Ceadwalla, delayed their death till he had converted them to Christianity. (Bede, *Hist. Eccl.*, tom. i., lib. iv., cap. xvi., p. 284, published by the English Historical Society.) All these places, with the exception of Redbridge, were more or less afforested. The district, however, seems to have been by far the most flourishing of any adjoining the New Forest, owing, no doubt, to the immigration which the various creeks invited, and the remains of salterns still show its former prosperity. Next to it came the Valley of the Avon, its mills often assessed, in *Domesday*, by a payment of the eels caught in the river.

3. Colonel Hammond, Governor of the Isle of Wight, in a letter to the Committee of Derby House, dated from Carisbrook Castle, June 25th, 1648, speaks of "Caushot Castle as a place of great strength." (Peck's *Desiderata Curiosa*, vol. ii, book ix., p. 383.) In the reign of Elizabeth there were stationed here a captain, with a fee of one shilling a day; a subaltern with eightpence; four soldiers and eight gunners with sixpence each; and a porter with eightpence. (Peck's *Desiderata Curiosa*, vol. i., book ii., p. 66.) And

in 1567, we find the queen ordering "the mountyng of ordinance," probably to pay attention to Philip, who was expected to pass through "the narrowe seas." Record Office. Domestic Series, No. 43, Aug. 27. 1567, f. 52.
4. *The Chronicle*. Ed. Thorpe. Vol. i. p. 24. *Florence*. Ed. Thorpe. Vol. i. pp. 3, 4.
5. Compare his edition of *The Chronicle*, vol. ii. p. 13, with note 1 at p. 4, vol. i., of *Florence*.
6. Early English Settlements in Great Britain—*The Proceedings of the Archaeological Institute*, the Salisbury volume, pp. 56-60. It is, of course, not without much consideration that I presume to differ from Dr. Guest; but surely the passages quoted from Bede refer to nearly 200 years after the arrival of Cerdic and his nephews, Stuf and Wihtgar, when their descendants would have been sure to have crossed over, finding the east side far richer than the cold, barren district where the New Forest afterwards stood.
7. *The Early and Middle Ages of England*, p. 56, foot-note. I may, perhaps, add, that Camden also placed it at Yarmouth; Carte, at Charmouth, in Dorsetshire; and Milner, at Hengistbury Head. Gibson, with some others, in his edition of *The Chronicle* (under nominum locorum explicatio, pp. 19, 20), alone seems to have fixed on this spot. Lappenburg, however, says that the site is no longer known. *England under the Anglo-Saxon Kings*. Ed. Thorpe, p. 107.
8. In a letter of Southampton's to Cromwell, 17th September, 1539 (*State Papers*, vol. i. p. 617), it is called Calsherdes; whilst in another letter of his, also to Cromwell (*Ellis's Letters*, second series, vol. ii. p. 87), he writes Calshorispoynte. Leland, in his *Itinerary* (Ed. Hearne, second edition, vol. iii., p. 94, f. 78), speaks of both "Cauldshore" and "Caldshore Castelle;" and again (p. 93, f. 77), calls it Cawshot, as it is also spelt in Baptista Boazio's Map of the Isle of Wight, 1591; whilst in the State papers of Elizabeth we find Calshord. (Record Office. Domestic Series, No. 43, f.

52. Aug. 27th, 1567.) I give these examples to show the number of variations through which the name has passed. No form is too grotesque for a corruption to assume. How names become corrupted, let me give an example in the word Hagthorneslad (from the Old-English "hagaþorn," a hawthorn), as it is written in the perambulation of the Forest in the twenty-ninth year of Edward I., which in Charles II.'s time is spelt Haythorneslade, thus losing its whole significance, although to this day the word "hag" is used in the Forest for a "haw," or "berry."
9. The simple termination "ore"—"ora," and not "oar," as spelt in the Ordnance Map, may be found within a stone's throw of Calshot, in Ore Creek.
10. See previously, chapter iv. p. 40, foot-note.
11. At the date of the Dauphin's leaving England, William de Vernon was dead, which makes his embarkation at Leap less probable. Neither Roger of Wendover (vol. iv. p. 32. Ed. Coxe), nor Walter Hemingburgh (vol. i. p. 259. Ed. Hamilton), nor Ralph Coggeshale (*Chronicon Anglicanum. Bouquet. Recueil des Historiens des Gaules et de la France*, tom, xviii. p. 113 C), nor the *Chronicon Turonense* (in the *Veterum Scriptorum Amplissima Collectio* of Marténe and Durand, tom. v. p. 1059 B), nor Rymer's *Foedera* ("De salvo conductu Domini Ludovici," tom. i. p. 222), say anything of the place of embarkation.
12. I believe on that of the Oglander MSS. in the possession of the Earl of Yarborough, but which I have never seen. Neither the *Iter Carolinum, Herbert's Memoirs* (London, 1572, p. 38), Huntington's account (same volume, p. 160), *Berkeley's Memoirs* (second edition, 1702, p. 65), *The Ashbumham Narrative* (London, 1830, vol. ii. p. 119), nor Whalley's letter in Peck's *Desiderata Curiosa* (tom, ii., lib. ix, pp. 374, 375), nor Hammond's, in *Rushworth's Collection* (part iv., vol. ii., p. 874), mention the place, though the latter would seem to indicate that the King sailed direct from Tichfield to Cowes. Ashburnham and Berkeley had, we know

from Berkeley (*Memoirs*, same edition as before, p. 57) and Ludlow (*Memoirs*, 1771, p. 93), previously gone by Lymington to the Island.

13. As the passage is so important, I give it in full:— Ἀποτυποῦντες δ' εἰς ἀστραγάλων ῥυθμοὺς κομίζουσιν εἴς τινα νῆσον προκειμένην μὲν τῆς Βρεττανικῆς, ὀνομαζομένην δὲ Ἴκτιν· κατὰ γὰρ τὰς ἀμπώτεις ἀναξηραινομένου τοῦ μεταξὺ τόπου ταῖς ἁμάξαις εἰς ταύτην κομίζουσι δαψιλῆ τὸν καττίτερον. Ἴδιον δέ τι συμβαίνει περὶ τὰς πλησίον νήσους τὰς μεταξὺ κειμένας τῆς τε Εὐρώπης καὶ τῆς Βρεττανικῆς· Κατὰ μὲν γὰρ τὰς πλημμυρίδας τοῦ μεταξὺ πόρου πληρουμένου νῆσοι φαίνονται, κατὰ δὲ τὰς ἀμπώτεις ἀπορρεούσης τῆς θαλάττης καὶ πολὺν τόπον ἀναξηραινούσης θεωροῦνται χερρόνησοι.—Lib. v., cap. xxii., vol. i., p. 438. Ed. Dindorf. Leipsic, 1828-31. Pliny, as Wesseling remarks, in his note on this passage, quoted by Dindorf, vol. iv. p. 421, by some mistake, makes the Isle of Wight (Mictis) six days' sail from England. See Sir G. C. Lewis's *Astronomy of the Ancients*, chap, viii., sect. iii. p. 453.

14. As before, sect. iv. p. 462.

15. *Topographical Remarks on the South-Western Parts of Hampshire*, vol. ii. pp. 5, 6, 1793.

CHAPTER VI.

BEAULIEU ABBEY.

Arches of the Chapter House

I SHOULD trust that, on a fine day, twenty miles are not too much for any Englishman. If they are, and any one should think the walk along the coast too long, Beaulieu may be reached by going direct from Hythe, across Beaulieu Common. The moor stretches out on all sides, flushed in the summer with purple heather, northward to the Forest, southward to the cultivated fields round Leap and Exbury. Passing "The Nodes,"[1] the road runs quite straight to Hill Top, with its clump of firs, which we reached in the last chapter.

Down in the valley, hid from us by a turn in the road, lies Beaulieu. But a little farther on we reach part of the old Abbey walls, broken here and there, clustered with ivy, and grass, and yellow mullein, and white yarrow, whilst vine-clad cottages stand against its sides. The village is situated on a bend of the Exe, where, spanned by a bridge, the stream falls over the weir, formerly turning the old mill-wheel of the monks, and then, broadening with the tide, winds through meadows and thick oak copses down to the Solent.

Although far more beautifully situated, the Abbey is not nearly so well known as its own filial house at Netley, simply because more out of the way. For a moment let us give some account of its foundation, illustrating as it does both King John's cruelty and superstition. The story, as told by the monks, is that John, after various oppressions of the Cistercian Order, in the year 1204, convened their abbots to his Parliament at Lincoln. As soon as they came, he ordered his retainers to charge them on horseback. No one was found to obey such a command. The monks fled to their lodgings. That night the King dreamt he was led before a judge, who ordered him to be scourged by these very monks. The next morning John narrated his dream, which was so vivid that he declared he felt the blows when he awoke, to a priest of his court, who told him that God had been most merciful in thus simply chastising him in this world, and revealing the secrets of His will. He advised him at once to send for the abbots, whom he had so ill-treated, and to implore their pardon.[2]

Some truth, doubtless, underlies this story. Certain it is that in the same year, or the next, John founded the Abbey at

Beaulieu, then Bellus Locus, so called from its beauty, placing there thirty monks from St. Mary's, at Citeaux, endowing it with land in the New Forest, and manors, and villages, and churches in Berkshire; exempting it from various services and taxes and tolls; giving further, out of his own treasury, a hundred marks; and ordering all other Cistercian Houses to assist in the work. Not only did he do this, but he revoked his gift of the manor of Farendon, which, in the previous year, he had conditionally bestowed on some other Cistercian monks, and now transferred it to Beaulieu, making the House at Faringdon a mere offshoot from the larger building.[3] And the abbot designate repaid him in his life-time by accusing his enemy, Stephen of Canterbury, before the Pope, for treason, and causing him to be suspended.[4]

John died, and Henry III. not only confirmed the privileges, but granted several more in consideration of the great expense of the building, and Innocent III., gave it the right of a sanctuary. So the work proceeded. The stone was quarried principally from the opposite limestone-beds in the Isle of Wight; and was brought over, says tradition, curiously illustrating the vague notions of ancient geography, which we have seen in Diodorus Siculus,[5] in carts. Not, however, till 1249, some forty-five years after its foundation, was the monastery finished. Henry himself, and his Queen, and Richard, Earl of Cornwall, and a long train of nobles and prelates, came to its dedication on the feast of St. John; Hugh, the first abbot, spending no less than five hundred marks on the entertainment.[6]

So, at last, the good work was accomplished, and men came here and lived, taking for their pattern the holy St. Benedict, and finding the problem of life solved by daily prayer to heaven and labour on earth.[7]

Here, to its sanctuary, in 1471, after she had landed at Southampton on Easter Day, the very day of the battle of Barnet, fled the Countess of Warwick, wife of the King-maker, slain on that bloody field.[8] Here, too, in 1497, after having raised the siege of Exeter, and deserting his troops at Taunton, fled the worthless Perkin Warbeck, not only an impostor, but a coward, closely pursued by Lord Daubeny and five hundred men. Persuaded, however, by Henry VII.'s promises, he left his shelter only to become a prisoner in the Tower, and finally to expiate his deceit at Tyburn.

So years passed at the Abbey, the monks happy in saying their daily prayers, content to see the corn grow, and their vineyards ripen, and their flocks increase, knowing little of the troubles which raged in the outer world, save when some forlorn fugitive arrived. But even what is best becomes the worst. Time brought a change of spirit on all the monasteries. Long before the middle of the sixteenth century the stern earnestness of a former age had dwindled into effeminacy and sensuality. Piety had sunk into gross idolatry; and faith, amongst the laity, had been corrupted into credulity, and, with the priests, into hypocrisy. The greatest blessings had festered into curses. It was so, we know, through all England. And Beaulieu must suffer with the rest of the monasteries.

In 1587, the Abbey was dissolved, the last Abbot, Thomas Stephens, with twenty out of the thirty monks, signing the deed of surrender.[9] Stephens was pensioned off with a hundred marks; and some of the monks received various annuities and compensations for their losses. So fell the monastery of Beaulieu, and its stones went to build Henry VIII.'s martello tower at Hurst, and its lead to repair Calshot,[10] to fight against the very Power which had raised it to its glory.

Nothing could be more beautiful than its situation on the banks of the Exe, formed by the tide more into a lake than a river. On every side it was sheltered: on the north by rising ground and the woods of the New Forest, and on the east again by the Forest and more hills, from whence an aqueduct brought down the water for the use of the monks; and on the south and west all was guarded by the river.

To this day the outer walls are in places standing, with the water-gate covered with ivy. And inside is the abbot's house, placed amongst its own grounds, surrounded by elms. Above its doorway is cut a canopied niche, where stood the patron saint, the Virgin, and above runs the string-course, supported by its carved corbel-heads. But the whole building has been unfortunately defaced by a moat and turretted wall, built as a defence by one of the Montagues against French privateers, as also by the modernized windows.

Entering, we come into the guesten-hall, the *magna camera arcuata*, formerly hung with tapestry, where the minstrels entertained the guests with songs or tales. Like all the other rooms, it has been sadly modernized, though its fine groined roof, springing from four shafts on each side, and a lancet window in the east wall, still remain. Upstairs, too, is left some oak panelling of Henry VIII.'s time, of the linen pattern, but covered over with paint. Eastward, in the meadow adjoining, stands the dormitory, better known in the village, from its former occupants, as Burman's House. Passing through it, we suddenly come upon the green quadrangle once surrounded with cloisters, where the three arches leading into the chapter-house still remain. The black Purbeck marble shafts, and bands, and capitals, have, however, long since become weather-worn and decayed, though the Binstead and Caen stone still stands, here and there covered with ivy, crested with wall-flowers, and white and crimson pinks, and rusted with lichens.

In the chapter-house are strewed the broken pillars which supported the groined roof, and the broken stone-seats which ran round the inside, whilst on the floor lie a stone coffin and gravestones. To the north of it stand the ruins of the sacristy, which had an entrance from the south transept of the church, from which, also, a staircase led to the scriptorium.

Of the cloisters, the north alley is the most perfect, with its seven carols, where the monks sat and talked; whilst above project the corbels which carried the cloister-roof. Here and there, too, as at the two north doors leading into the church, some of the original pavement still remains, and at the south-east corner a staircase led to the lavatory.

The church, however, has long since been destroyed. Nothing, except a portion of the south transept, is left. The foundations, though, can be accurately traced, showing the nave and aisles, and the large circular apse at the east end. Scattered about, too, appear the tesselated floor, bright as on the day it was laid down, and the graves of the abbots, and of Eleanor of Acquitaine, mother of the founder.[11]

Out in the fields beyond stand the ruins of a building, now a mere pinfold for cattle, called by tradition the Monk's Vine-Press, whilst the meadows beyond, lying on the slope of the hill, are still known as "the Vineyards."[12]

But the refectory still remains on the south side of the cloisters, from which a doorway, still ornamented with iron scroll-work, used to lead. Ever since the Reformation it has served as the parish church, differing only in its appearance by its lack of orientation.[13] In 1746 it was repaired, and its original roof lowered, and its fine triplet at the south end spoilt by a buttress, and one of the lancets lighting the wall-passage on the west side also blocked up. Its walls, however, are now covered with common spleenwort, and wall-lettuce, and pellitory, whilst the narrow-leaved rue—the "herb o' grace o' Sundays"—with which the old churchyards used to be sown, shows its pale blue blossoms amongst the gravestones.

Pulpit of the Refectory.

Inside it is still more interesting. Here still stands the lovely stone pulpit, its panels rich with flower-tracery, approached by a wall-passage and open arcade springing from double rows of black Purbeck marble pillars. This was the old rostrum of the monks, where one of the brethren read to the rest at their meals; so that, as St. Augustine says, their mouths should not only taste, but their ears also drink in the Word of God. Here, in this very village church, the old Cistercian monks obeyed the injunction which the Bishop of Hippo gives to the canons of his own order,—"When we enter, let us bare our heads, and going to our seats bend before the cross. Let us not behave idly, lest we give offence to any one. Let not our eyes wander, lest we give occasion for bickering, or quarrelling, or laughing; but fulfilling the saying of the blessed Hugh of Lincoln, 'let us keep our eyes upon the table, our ears with the reader, and our hearts with God.'"[14]

In the churchyard, plainly traceable by the ruined foundations, and mounds, and depressions, are the sites of the lavatory and kitchens, whilst in the fields beyond lie the fish-ponds. Everywhere, in fact, are seen the traces of the monks. Their walks still remain by the side of the Exe, overgrown with oaks, bright in the spring with blue and crimson lungwort, and sweet with violets, such as grew when Anne Beauchamp sought refuge here that dismal Easter day.

Not only do the Abbey grounds,[15] but the whole district, show the size of the monastery. Going out of Beaulieu, upon the road to Bucklershard, we come upon the ox-farm of the monks, still called Bouvery, and still famous for its grazing land. A little further, about the centre of their various farmsteads, at St. Leonard's, better known now as the Abbey Walls, stands part of the large barn, or *spicarium*, of the monastery, such as still remains in other parts of England—at Cerne Abbey, and Abbotsbury, and Sherborne, and Battle Abbey.[16] A modern barn now stands within it, partly formed by its walls, but its original size is well shown by the lofty eastern gable, locally called the Pinnacle, which, covered with ivy, overhangs the road.

The Barn of St. Leonard's Grange

Close to the old farm-house, built from its ruins, stands a small roofless Decorated Chapel. The west window, and the arch of

The Chapel of St Leonard's Grange

the west doorway, still remain, and at the same end still project the corbels which supported the gallery. In the east wall are canopied niches, under which stood figures; and on the south the stoup, and the broken conduit where the holy water ran, and two aumbries are still visible, whilst opposite to them, in the present doorway, another aumbrie is inserted with its two grooves for shelves cut in the stone.[17]

Close to St. Leonard's lies also the sheep-farm of the monks, still called Bargery, and still famous for its sheep-land. Nearer Bucklershard is Park Farm, another grange, where fifty years ago stood a chapel, smaller even than the one we have just seen, partly Early-English and Decorated. It was divided into two compartments by a stone screen reaching to a plain roof. The piscina in the south wall was finally used by the ploughmen to mix their wheat with lime, until the whole building was pulled down to enlarge the farm-house from whose south-east end it projected.[18]

At these two granges the brethren worked in summer from chapter till tierce, and from nones till vespers. Here lived the ploughmen and artisans, the millers, and smiths, and carpenters, of the monastery. For them were these chapels built, lest either the weather or the roads might prevent them going to the Abbey Church.[19] Here they all worshipped as one family, the serf no longer a serf, but a freedman, when he entered the service of the abbey.

Farther away to the westward lies Sowley Pond, called in the Abbey Charters Colgrimesmore, and Frieswater, covering some ninety acres, formerly the boundary of the abbey estate, and used by the monks as a preserve for their fish. Here once were iron-works, whose blast-furnaces were heated with wood and charcoal from the Forest. The iron-stone was brought from Hengistbury Head and the Hordle Cliffs, and after being melted was shaped by the tilt-hammers, and finally sent off inland to Reading, or shipped at Pitt's Deep. But like all the other ferraria of Sussex and Hampshire, these too have long since been stopped, driven out of the field by the Staffordshire iron-works. Nothing now remains to tell their former importance but a few mounds and the village Forge-Hammer Inn, and a country proverb, "There will be rain when Sowley hammer is heard," whose meaning is fast being lost.

Returning, however, to Beaulieu, let us once more look at the old abbey and the ruins of the cloisters, and try to imagine for ourselves the time when, secluded from the world, in the midst of the New Forest, the monks from Citeaux prayed and worked, clad in their coarse white woollen robes, and slept, according to their vow, on pallets of straw, giving shelter to the fugitive, and food to the hungry.[20] It is only by seeing some such grey ruins as these, still breathing of a long past religion, placed amongst the solitude of their own green meadows and woods, by the silent lapse of some stream flowing and ebbing with every tide, that we can at all understand the meaning of a life of contemplation, and its true value. Along these cloisters paced the brethren, their eyes bent on the earth, their thoughts on heaven. Here tolled the great abbey-bell, its sound, full of solemn sweetness, borne not only over the lonely Forest, but down the river seaward to the tossing sailor. Here was that comfort, which could never fail, offered to the most desolate, and heaven itself, as a fatherland, to the exile. Here the great gate not only rolled back the noise of the world, but, to show that mercy is ever better than vengeance, stayed the hand of the law, and blunted the sword of the pursuer.

In these days we are surrounded by noise and excitement. Everywhere is haste and its accompanying confusion. It matters not what we do, the fever of competition ever rages. We travel as though we were flying from ourselves. We write the history of things before they are accomplished, and the lives of men before they are dead. Sorely there is some profit to be found in coming to a quiet village like this, if it will only give us some glimpses of a life which stands out in such strange contrast to our own.

Canopied Niche in St. Leonard's Chapel.

Footnotes

1. For an account of the barrows on Beaulieu Heath, see ch. xvii.
2. Dugdale's *Monasticon Anglicanum*. Ed. 1825, vol. v., p. 682. Num. ii. See *Chronica de Kirkstall*. Brit. Mus. Cott. MSS. Domitian. A. xii., ff. 85, 86. The cause of John's enmity against the Cistercian Order may be gathered from Ralph Coggeshale, *Chronicon Anglicanum*, as before in Bouquet, *Recueil des Historiens des Gaules et de la France*, tom, xviii. pp. 90, 91.
3. *Carta Fundationis per Regem Johannem*, given in Dugdale (Ed. 1825, vol. v. p. 683); and *Confirmacio Regis Edwardi tertii super*

cartas Regis Johannis, Brit. Mus., Bib. Cott. Nero, A. xii., No. v, ff. 8-15, quoted in Warner (*South-West Parts of Hampshire*, vol. ii., Appendix, pp. 7-14). There are, however, no less than three dates given for its foundation. The *Annals of Parcolude*, according to Tanner (*Notitia Monastica*, Ed. Nasmyth, Hampshire, No. vi. foot-note h), say 1201, which is manifestly wrong; whilst John de Oxenedes, better known as St. Benet of Hulme (*Chronica*, Brit. Mus., Bib. Cott., Nero, D. ii., f. 223 K), with the *Chronicon de Hayles and Aberconwey* (Brit. Mus., Harl. MS., No. 3725, f. 10), and Matthew Paris, according to Dugdale, say respectively 1204 and 1205, though I have not been able to verify the last reference.
4. *Roger of Wendover*, English Historical Society. Ed. Coxe, vol. iii. p. 344.
5. See the previous chapter, pp. 57, 58, foot-note.
6. Curiously enough, as Warner remarks (vol. i. 267), Matthew Paris gives two dates for the dedication, the first 1246 (*Hist. Angl.*, tom. i. p. 710, Ed. Wats., London, 1640); and the second (p. 770) 1249; not, however, 1250, as Warner says, and who, followed by all later writers, totally misunderstands the passage, which means that, although the abbot spent so large a sum, yet the King would not remit him the fines he had incurred by trespass in the Forest,—"Nec tamen idcirco aliquatenus pepercit rex, quin maximum censum solveret illi pro transgressione quam dicebatur regi fecisse in occupatione Forestæ."
7. See Matthew Paris, in praise of the Cistercian Order. Same edition as before, tom. i. p. 916.
8. Not Margaret of Anjou, as the common accounts say, who, landing at "Weymouth, took refuge at Cerne Abbey. See *Historie of the Arrival of Edward IV. in England*, pp. 22, 23, printed for the Camden Society, 1838; and Hollinshed's *Chronicles*, vol. iii. p. 685; and Speed, B. ix. p. 866. Hall, however (*The Union of the Families of Lancaster and York*, p. 219), with Grafton, in his prose continuation of Hardyng (Ed. Ellis, 1812, p. 457), say it was to

Beaulieu that Margaret fled. But they are evidently mistaken, as Speed and Hollinshed, and the explicit and circumstantial narrative of the author of the *Historie*, show.

9. The following list of books at Beaulieu, quoted, with some omissions, by Warner (vol. i. p. 278) from Leland (*Collect. de Rebus Brit.*, vol. iv. p. 149), taken just before the dissolution, will show what was in those days an average ecclesiastical library:—

The Life of Archbishop Anselm, by Edmerus the monk, bound up with the *Life of Bishop Wilfrid*; *Stephanas on Ecclesiasticus*; *Stephanus on the Book of Kings*; *Stephanus on the Parables of Solomon*; *John, Abbot of Ford, on the Canticles*; *Damascenus on the Acts of Balaam and Josaphat*; a small book of *Candidas Arrian*; a small book of *Victorinus, the Rhetorician, against Candidas*; three books of *Claudian, respecting the Stale of the Soul, to Sidonius Apollinaris*; *Gislebertus on the Epistles of St. Paul*; *Prosper on a Life of Contemplation and of Activity.*

10. *Ellis's Letters*, second series, vol. ii. p. 87. For Henry VIII.'s enforcement of Wolsey's levies on Beaulieu, see *State Papers*, vol. i., part ii., p. 383.

11. A few years ago, when the foundations were being excavated, a female skeleton was found near the high altar, and was supposed, by its position, to have been that of the Queen.

12. Warner (vol. i. 255) mentions that in his time there was still brandy in the steward's cellars made from the vines growing on the spot. *Domesday* gives several entries of wines (see Ellis's *Introduction*, vol. i. pp. 116, 117), though none in the Forest district But the term 'Vineyards' is still frequently found hereabouts as the name of fields generally marked by a southern slope, as at Beckley and Hern, near Christchurch, showing how common formerly was the cultivation of the vine, first introduced into England by the Romans.

13. In Brit. Mus., Harl. MS. 892, f. 40*b*, is an extract from a most interesting letter written in 1648, describing the state of the

refectory, which seems, with the exception of the alterations made in 1746, to have been much the same as at present.

14. Quoted from Dugdale's *Monasticon Anglicanum* by Warner, vol. i. p. 249.

15. It is pleasant to have to add that the present noble owner, the Duke of Buccleuch, has shown not only good taste and judgment in the restoration of the dormitory and the excavation of the church, but a wise liberality in throwing the grounds open to the public.

16. In Parker's *Glossary of Architecture* is given a list of some of these old barns. Vol. i. pp. 240, 241.

17. Some curious leaden pipes, soldered only on one side, were dug up close by, which are worth seeing, as they show how late the process of running hollow lead pipes was invented. The earthenware pipes found with them are as good as any which are now made. At Otterwood Farm, on the other side of the Exe, pavement and tiles have also been discovered.

18. The chapel was standing in Warner's time. *South-Western Parts of Hampshire*, vol. i. pp. 232, 233.

19. In Brit. Mus., Bib. Cott, Nero, A. xii., No. vii. f. 20 a b, is a copy of a Bull from Alexander I., giving permission to all the Cistercian Houses to hold service at their granges.

20. Even Layton saw their kindness, and pleaded for the poor wretches whom they had protected. Letter regarding Beaulieu Sanctuary from Layton to Cromwell, *Ellis's Letters*, third series, vol. iii. pp. 72, 73.

CHAPTER VII.

THE SOUTH-WESTERN PART—BROCKENHURST, BOLDRE, SWAY, HINCHELSEA, AND BURLEY.

View in Frame Wood, Brockenhurst.

At present we have seen nothing of the actual Forest. It is only as we go northward that we begin to enter its woods. Instead of the old Forest track, a road now runs from Beaulieu to Brockenhurst, along which we will go. So, leaving the village, and passing a few straggling half-timbered cottages, we reach Stickland's Hill, where, down in the valley, we can see the Exe winding round the old Abbot's House set amongst its green elms. Farther on we come to Hatchet Gate, and the Forest then spreads before us, with Hatchet Pond on our left, and Little Wood and the Moon Hill Woods on our right; whilst, here and there on the common, rise scattered barrows.

And now, instead of keeping to the road, let the reader make right across the plain, by one of the Forest tracks, to the woods at Iron's Hill. The stories, with which most books on the Forest abound, of persons being swamped in morasses, are much exaggerated. Mind only this simple rule—wherever you see the white cotton-grass growing, and the bog-moss particularly fine and green, to avoid that place.

And now, when you are fairly out on the moor, you will feel the fresh salt breeze blowing up from the Solent, and see the long treeless line of the Island hills in strange contrast with the masses of wood in front; whilst the moor itself, if it be August, waves with purple and crimson, except where, here and there, rise great beds of fern—green islands, in the red sea of heath.

Most of the finest timber at Iron's Hill and Palmer's Water has been lately cut. Keeping on, however, we shall again come out upon the road which leads down to the stream, close to a mill. Passing over the footbridge, we skirt Brockenhurst Manor, where, at Watcombe, once lived Howard the philanthropist, and so at last reach the village.

So greatly has the Forest been reduced in size, that Brockenhurst, once nearly its centre, is now only a border village. Its Old-English name (the badger's wood), like that of Everton the wild-boar place, on the southern side of the Forest, tells its own story. It consists of one long straggling street, and a few scattered houses, with one or two village inns. Much of its wildness has been spoilt by the railroad; and in consequence, too, of the adjoining manor of Brockenhurst, it appears even less than it really is in the Forest. Still, however, if the reader wishes to see the Forest woods and heaths towards the south, let him come here, and the village accommodation will only give an additional charm to the scenery. I, for my part, do not know that a clean English village inn, with its sanded floor, and its best parlour kept for state occasions, makes such bad quarters. It is a real pleasure to find some spots on the earth not yet disfigured by fashionable hotels.

At Brockenhurst existed one of those tenures of knight-service once so common throughout England. Here Peter Spelman, an ancestor of the antiquary, held a carucate of land by the service of finding for Edward II. an esquire clad in coat of mail for forty days in the year, and whenever the King came to the village to hunt, litter for his bed, and hay for his horse, which last clause will give us some insight into the often rough living and habits of the fourteenth century.[1] Here, too, not many years ago, droves of deer would at night, when all was still, race up the village street, and the village dogs leap out and kill them, or chase them back to the Forest.

The church, one of the only two in the Forest mentioned in *Domesday*,[2] is built on an artificial mound on the top of a hill, a little way out of the village, so that it might serve as a landmark in the Forest.[3] The church has been sadly mutilated. A wretched brick tower has been patched on at the west end; and on the north side a new staring red brick aisle, which surpasses even the usual standard of ugliness of a dissenting chapel. On the south side stands the Norman doorway, with plain escalloped capitals, and an outside arch ornamented with the indented and chevron mouldings. The chancel is Early-English, whilst the plain chancel arch which springs without even an impost from the wall, is very early Norman. Under one of the chancel windows rises the arch of an Easter sepulchre, whilst a square Norman font, of black

Purbeck marble, stands at the south-west end of the nave.

If the church, however, has been disfigured, the approach to it fortunately remains in all its beauty. For a piece of quiet English scenery nothing can exceed this, A deep lane, its banks a garden of ferns, its hedge matted with honeysuckle, and woven together with bryony, runs, winding along a side space of green, to the latch-gate, guarded by an enormous oak, its limbs now fast decaying, its rough bark grey with the perpetual snow of lichens, and here and there burnished with soft streaks of russet-coloured moss; whilst behind it, in the churchyard, spreads the gloom of a yew, which, from the Conqueror's day, to this hour, has darkened the graves of generations.[4]

But the charm of Brockenhurst, as of all the Forest villages, consists in the Forest itself. To the north runs the small Forest stream, blossomed over in the summer with water-lilies. On the left lies Black Knoll, with its waste of heath and gorse, running up to the young plantations of New Park. On the right, Balmor Lawn, with its short, sweet turf, where herds of cattle are pasturing, stretches away to Holland's Wood, with old thorns scattered here and there, in the spring lighting up the Forest with their white may.

Just now though, it is the southern part of the Forest we must see. So going back again for a little way upon the Beaulieu Road, and leaving it just above the foot-bridge for Whitley Lodge, let the reader go on to Lady Cross. Suddenly he will come out upon the northern edge of Beaulieu Heath, and see again the Island Hills. To the people in the Forest, the Island is their weather-glass. If its hills look dark blue and purple, then the weather will be fine; but if they can see the houses and the chalk quarries on the hill sides, the rain is sure to come.

Keeping straight on, with Lady Cross Lodge to our left, we enter Frame Wood, with its turf and its bridle roads winding under the Forest boughs. Down in the bottom runs the railroad bending away to the north. On the other side, the thick woods of Denny rise; and the clump of solitary beeches on the top of the knoll shows the last remains of Wood Fidley, so well known as having given rise to the Forest proverb of "Wood Fidley rain," that is, rain which lasts all the day.

Here you can wander on for miles, as far as the manor of Bishop's Ditch, belonging to Winchester College, which the Forest peasant will tell you was a grant of land as much as the Bishop of Winchester could in a day crawl round on his hands and knees. As to losing yourself, never mind. The real plan to enjoy the Forest is to wander on, careless whether you lose yourself or not. In fact, I believe the real method is to try and lose yourself, finding your greatest pleasure in the unexpected scenes of beauty into which you are led.

There are plenty of other Forest rambles round Brockenhurst which must not be forgotten. Just at the western edge of Beaulieu Heath, about three miles off, stands Boldre Church, with its solitary churchyard surrounded by trees. On one side, it looks out upon the bare Forest; and on the other, down into the cultivated valley. Most suggestful, most peaceful is this twofold prospect, telling us alike both of work and companionship, as, too, of solitude—all of them, in religion, so needful for man. Its tower stands boldly out, almost away from the church, just between the nave and the chancel, serving formerly, like Brockenhurst steeple, as a landmark to the Forest; whilst the long outline of the nave is broken only by the south porch, and its three dormer windows. Close to the north side, under the shadow of a maple, lies one of the truest lovers of Nature—Gilpin, the author of the *Scenery of the New Forest*, with a quaint, simple inscription on his gravestone written by himself. In the church are tablets both to him and Bromfield, the botanist, a man like him in many ways, but who, dying abroad, was not allowed to rest beside him in this quiet graveyard.

Here, too, Southey married his second wife, Caroline Bowles. The south aisle is the oldest part, with its three Norman arches rising from square piers, whilst the north aisle is divided from the nave by a row of Early-English arches springing from plain black Purbeck marble shafts. In the east window of this aisle were once painted the arms of the Dauphin of France—the fleurs de lis—blazoned, as they were formerly, over the whole field, telling us the story of Lewis having been invited to England and crowned king by John's barons, and whose traditional flight at Leap has been mentioned.

Down below, in the valley of the Brockenhurst Water, lies Boldre, the Bovre of *Domesday*, with its meadows and cornfields. It is worth while to pause for a moment, and notice the corruption of Boldre into Bovre by the Norman clerks. The word is from the Keltic, and signifies the full stream ("*y Byldwr*"), and has nothing to do with oxen. We must, too, bear in mind that the various Oxenfords and Oxfords are themselves corruptions, and really come not from oxen at all, but Usk, literally meaning the stream-ford or stream-road, and are in no way connected with the various Old-English Rodfords to be found in different parts of the kingdom. This corruption of language we see daily going on in our own Colonies, but it is well to pause and remember that the same process has taken place in our own country.

Passing over the bridge, and up the village, and under the railway arch, we once more reach the Forest at Shirley Holms, coming out on Shirley and Sway Commons. Here again, as on Beaulieu Heath, there is not a single tree, nothing but one vast stretch of heather, which late in the summer covers the ground with its crimson and amethyst. There is only one fault to be found with it, that when its glory is past it leaves so great a blank behind: its grey withered flowers and its grey scanty foliage forming such a contrast with its previous brightness and cheerfulness.

But these two commons will at all times be interesting to

the archaeologist and historian. On the north-east side lies the Roydon, that is, the rough ground, a word which we find in other parts of the Forest; and not far from it is Lichmore or Latchmore Pond—the place of corpses—which is confirmed by the various adjoining barrows.[5]

After this point, there is nothing to attract the traveller, unless he is a botanist, to the south. Wootton, and Wilverley, and Setthorns, and Holmsley, are all young plantations, whilst at Wootton the Forest now entirely ceases, though once stretching five miles farther, as far as the sea. So let him make his way to Longslade, or Hinchelsea Bottom, as it is indifferently called, where about the middle of June blossoms the lesser bladderwort (*Utricularia minor*), and about the same time, or rather later, the floating bur-reed (*Sparganeum natans*).

Above, rises Hinchelsea Knoll, with its old hollies and beeches; and still farther to the north the high lands round Lyndhurst and Stony Cross crowned with woods. Westward, the heather stretches over plain and hill till it reaches Burley, Making right through Hinchelsea, and then skirting the north side of Wilverley plantations, we shall come to the valley of Holmsley, so beloved by Scott, and which put him in mind of his native moors, without seeing which once a year, he so pathetically said, he felt as if he should die. Its wild beauty, however, is in a great measure spoilt by the railroad, and the large trees which grew in Scott's time have all been felled.

Burley itself, which now lies just before us, is one of the most primitive of Forest hamlets, the village suddenly losing itself amongst the holms and hollies, and then reforming itself again in some open space. So thoroughly a Forest village, it is proverbially said to be dependent upon the yearly crop of acorns and mast, or "akermast," as they are collectively called. To the south-west stands Burley Beacon, where some entrenchments are still visible, and the fields lying round it are still called "Greater" and

"Lesser Castle Fields," and "Barrows," and "Coffins," showing that the whole district has once been one vast battle-field.

Close to the village are the Burley quarries, where the so-called Burley rock, a mere conglomerate of gravel, the "ferrels," or "verrels," of North Hampshire, is dug, formerly used for the foundations of the old Forest churches, as at Brockenhurst, and Minestead, and Sopley in the Vale of the Avon. The great woods round Burley have all been cut, except a few beech-woods, but here and there "merry orchards" mingle themselves with the holms and hollies, wandering, half-wild, amongst the Forest.[6]

Turning away from the village, and going north-east, before us rise great woods—Old Burley, with its yews and oaks, where the raven used to build; Vinney Ridge, with its heronry at one extremity, and the Eagle Tree at the other; whilst behind us are the young Burley plantations. Here, near the Lodge, scattered in some fields, stand the remains of the "Twelve Apostles," once enormous oaks, reduced both in number and size, with

> "Boughs moss'd with age,
> And high tops bald with dry antiquity."

And now, if the reader does not mind the swamps—and if he really wishes to know the Forest, and to see its best scenes, it is useless to mind them—let him make his way across to Mark Ash, the finest beechwood in the Forest, which even on a summer's day is dark at noon. Thence the wood-cutter's track will take him by Barrow's Moor and Knyghtwood, where grows the well-known oak. Here a different scene opens out with broad spaces of heath and fern, where the gladiolus shows its red blossoms among the green leaves of the brake; whilst on the hill, distinguished by its poplars, stands Rhinefield, with its nursery, and, below, the two woods of Birchen Hat, where the common buzzard yearly breeds.

Keeping along the main road, which is just before us, nearly as far as the New Forest Gate, we will turn in at Liney Hill Wood, going through the woods of Brinken, and the Queen's Mead, and the Queen's Bower, following the course of the stream.

Very beautiful is this walk, with its paths which stray down to the water's edge, where the cattle come to drink; the stream pausing round some oak roots, which pleach the banks, lingering in the darkness of the shade, and at last going away with reluctance.

Few things, of their sort, can equal these lowland Forest streams, the water tinged with the iron of the district, flashing into amber in the sunlight, and deepening into rich browns in the shade, making the pebbles hazel as it ripples over them.

All the way along grow oaks and beeches, each guarded with its green fence of kneeholm, and furred with moss, which the setting sun paints with bands of light. And so, in turn, passing Burley and Rhinefield Fords, and Cammel Green, and the Buckpen, where the deer used to be fed in winter, the path suddenly comes out by a lonely grass-field, known as the Queen's Mead, and immediately after enters the Queen's Bower Wood. At the farther end, a bridge crosses the brook by the side of one of the many Boldrefords in the Forest; and in the distance, across Black Knoll, shine the white houses of Brockenhurst.

View in the Queen's Bower Wood

Footnotes

1. Blount's *Fragmenta Antiquilatis*. Ed. Beckwith, p. 80, 1815. *Testa de Nevill*, p. 235 a (118). We know, however, that our forefathers, long before this, possessed beds, or rather cots, hung round with rich embroidered canopies. For their general love, too, of comfort and personal ornament and dress, we need go no further than to Chaucer's description of "Richesse," in his *Romaunt of the Rose*. Englishmen, however, were still then, as now, ever ready to lead a rough life if necessary, and to make their toil their pleasure.
2 In that portion of it which comes under the title of "In Foresta et

circa eam." See chap. iii. p. 31.
3. All over England did the church towers serve as landmarks, alike in the fen and forest districts. Lincolnshire and Yorkshire can show plenty of such steeples. At St. Michael's at York, to this hour, I believe, at six every morning, is rung the bell whose sound used to guide the traveller through the great forest of Galtres; whilst at All Saints, in the Pavement, in the same city, is shown the lantern, which every night used to serve as a beacon.
4. The following measurements may have some interest, and can be compared with those of the oaks and beeches in the Forest, given in chap. ii. p. 16, foot-note:—Circumference of the oak, twenty-two feet eight inches. Yew, seventeen feet. An enormous yew, completely hollow, however, stands in Breamore churchyard, measuring twenty-three feet four inches. There are certainly no yews in the Forest so large as these; and their evidence would further show that at all events the Conqueror did not destroy the churchyards. As here, too, there remains some Norman work in the doorway of Breamore church.
5. For some account of these barrows, see chapter xvii.
6. The word is from the French *merise*. At Wood Green, in the northern part of the Forest, a "merry fair" of these half-wild cherries is held once a week during the season, probably similar to that of which Gower sung.

CHAPTER VIII.

THE CENTRAL PART—LYNDHURST.

The Great Huntley Woods

As we leave Brockenhurst we find ourselves more and more in the Forest. The road to Lyndhurst is one long avenue of trees—beeches with their smooth trunks, oaks growing in groups, with here and there long lawns stretching far away into distant woods. Most beautiful is this road in the spring. Stand on the top of Clay Hill, about the beginning or end of May, and you shall see wood after wood, masses of colour, the birches hung with the softest green, and the oak boughs breaking into amber and olive, made doubly bright by the dark gloom of firs, the blackthorn giving place to the sweeter may, and the marigold on the stream to the brighter lily.

On our left lies New Park, now turned into a farm, where in 1670 Charles II. kept a herd of red deer, brought from France, but previously used as a pound for stray cattle. Passing on by a roadside inn with the strange sign of the "Crown and Stirrup," referring to a pseudo-relic of Rufus's, preserved at the King's House, but which is nothing more than a stirrup-iron of the sixteenth century, we reach Lyndhurst—the lime wood,[1] the capital of the Forest, the Linhest of *Domesday*.

William the Conqueror at one time held the place, which was once dependent on the royal manor of Amesbury. Here, after the afforestation, Herbert the Forester held one yardland, on which only two borderers lived, the rest of the manor, which was only two hydes, being thrown into the Forest.[2] Here, also, as at Brockenhurst, was another of the old feudal tenures; for, in the time of Edward I., William-le-Moyne held probably these same two hydes of land, which had been disafforested, by the sergeantry of keeping the door of the King's larder.[3]

In the village stands the Queen's House, built in Charles II.'s time, and adjoining it is the Hall where the Courts of Attachment, or Woodmote, the last remnant of the terrible Forest Laws, are regularly held by the verderers, to try all cases of stealing fern and timber.

Close by is the new, half-finished, church, standing in the old churchyard made famous by Mr. Kingsley's ballad. It is not fair at present to pass a final judgment. When the tower is added, and time shall have touched the walls with a soberer tone, its two great defects will have disappeared, though nothing can remedy the heavy and poverty-stricken window of the north transept with its flattened mullions, and a wretched chimney near the choir utterly spoiling the effect of the beautiful chancel windows.

Inside, the red brick pillars of the arches of the aisles are clustered with black slate shafts, and banded with scroll-work of white Caen stone, the capitals carved with lilies, and primroses, and violets.[4] And above hangs a Perpendicular timbered roof, resting on the corbel heads of the martyrs and reformers of the Church—of Melancthon and Cranmer, and Luther and Latimer,—and the carved emblems of the Evangelists at the four corners.

In the choir and chancel the wall-colouring is more harmonious than in the nave, where there is a certain coldness and hardness, whilst the shafts are here wrought of rich Cornish marble. Over the communion-table is Mr. Leighton's fresco, a small piece of it now only completed—an angel standing with outstretched hands, keeping back those virgins who have come too late to the bridegroom's feast, the despair and anguish of their faces further typified by the rent wall, and the melancholy dreariness of the owl.[5]

That there are defects in the church its greatest admirers would admit—the poorness of the roof, the harshness produced by the introduction of so much white, as also the bad colour of the bricks, and a heaviness which hangs over the clerestory windows of the nave. But, on the whole, it stands as a proof of the great advance during the last ten years of Art, as a cheering sign, too, that, amidst all the failures of Government, some taste and zeal are to be found amongst private persons.

There is nothing else of interest in the village. Once here busy scenes must have taken place, when the King came to hunt with his retinue of nobles; when down the street poured the train of bow-bearers, and foresters, and keepers, clad in doublets of Lincoln green, holding the dogs in leash. Then the woods rang with the notes of the bugle, and the twang of the bow-string sounded as the bolt, or the good English yard-shaft, brought down the quarry. Here, too, in the Civil War, were quartered grim Puritan soldiers, and prayers took the place of feasts.[6] Now, all is quiet. Nothing is to be seen but the Forest inviting us into its green glades.

The people of Lyndhurst ought, I always think, to be the happiest and most contented in England, for they possess a wider park and nobler trees than even Royalty. You cannot leave the place in any direction without going through the Forest. To the east lie the great woods of Denny and Ashurst; and to the north rise Cutwalk and Emery Down, looking across the vale to Minestead, and below them Kitt's Hill, and the woods stretching away towards Alum Green. On the extreme west Mark Ash, and Gibb's Hill, and Boldrewood, rise towering one after the other; whilst to the south stretch Gretnam and the Great and Little Huntley Woods, which the Millaford Brook skirts, here and there flowing out from the darkness of the trees into the sunshine, the banks scooped into holes, and held together only by the rope-work of roots.

These woods are always beautiful. Of their loveliness in spring we have spoken; and if you come to them in summer, then the first purple of the heather flaunts on every bank, and edges the sides of the gravel-pits with a crimson fringe; and the streams now idle, suffer themselves to be stopped up with water-lilies and white crowfoot, whilst the mock-myrtle dips itself far into the water. Then is it you may know something of the sweetness and the solitude of the woods, and wandering on, giving the day up to profitable idleness, can attain to that mood of which Wordsworth constantly sings, as teaching more than all books or years of study.

The Woodman's Path, Bramble Hill.

Footnotes

1. An objection, that the lime-tree was not known so early in England, has been taken to this derivation. This is certainly a mistake. In that fine song of the Battle of Brunanburh, we find—
"Bordweal clufan·
Heowan heaþolinde·
Hamora lafan."
(*The Chronicle*. Ed. Thorpe, Vol i. p. 200.)
The "geolwe lind" was sung of in many a battle-piece. Again, as

Kemble notices (*The Saxons in England*, vol. i., Appendix A, p. 480), we read in the *Cod. Dip.*, No. 1317, of a marked linden-tree. (See, also, same volume, book i., chap, ii., p. 53, foot-note.) Then, too, we have the Old-English word *lindecole*, the tree being noted for making good charcoal, as both it and the dog-wood are to this day. Any "Anglo-Saxon" dictionary will correct this notion, and names of places, similarly compounded, are common throughout England.

2. The entry in *Domesday* (facsimile of the part relating to Hampshire, photo-zincographed at the Ordnance Survey, 1861, p. iv. a) is as follows:—"In Bovere Hundredo. Ipse Rex tenet Linhest. Jacuit in Ambresberie de firma Regis. Tunc, se defendebat pro ij hidis. Modo, Herbertus forestarius ex his ij hidis unam virgatam (tenet), et pro tanto geldat; aliæ sunt in foresta. Ibi modo nichil, nisi ij bordarii. Valet x solidos. Tempore Regis Edwardi valuit vi. libras." It is worth noticing that Lyndhurst is here put by itself, and not with Brockenhurst and Minestead, and other neighbouring places under "In Nova Foresta et circa eam;" a clear proof, which might be gathered from other entries, that the survey was not completed.

3. Blount's *Fragmenta Antiquitatis*. Ed. Beckwith, p. 183. 1815. Here the place is called Lindeshull.

4. Let me especially call attention to the exquisite carving of some thorns and convolvuluses in the chancel. It is a sad pity that this part of the church should be disfigured by glaring theatrical candlesticks and coarse gaudy Birmingham candelabra.

5. I have only seen but the slightest portion of this fresco, so that it is impossible to properly judge of even the merits of this part. No criticism is true which does not consider a work of Art as a whole. At present, the angel with outstretched hands, full of nervous power and feeling, seems to me very admirable, though the position and meaning of the cloaked and clinging figure below is, at the first glance, difficult to make out; but this will doubtless,

as the picture proceeds, become clear. The richness, however, of the colouring can even now be seen under the enormous disadvantage of being placed beneath the strong white glare of light which pours in from the east window. Further, Mr. Leighton must be praised for his boldness in breaking through the old conventionalities of Art, and giving us here the owl as a symbol of sloth, and the wretchedness it produces.

6. Herbert's *Memoirs of Charles I.*, p. 95.

CHAPTER IX.

MINESTEAD AND RUFUS'S STONE.

Oaks in Boldrewood.

ABOUT four miles off from Lyndhurst lie Minestead and Rufus's Stone. There are three or four different roads to them. The most beautiful, though the longest, is over Emery Down, where, turning off to the left, you pass the woods of Kitt's Hill, and James Hill. Then crossing Millaford Bridge, and skirting on each side of the road the beeches of Holme Hill, and passing through Boldrewood, you make your way eastward across the stream below the Withy Bed Hat, and go through the woods of Puckpits and Stonehard.

Another road to the Stone is through Minestead by a footpath which crosses Mr. Compton's park, dotted with cottages, each with its garden full in the summer and autumn of flowers—yellow Aaron-rods, pink candy-tufts, colchicums, and marigolds, and tall sheaves of grey Michaelmas daisies.

In the village stands "The Faithful Servant," copied from the well-known picture at Winchester College. A little farther on we ascend Stoneycross Hill, the village orchards full of Mary-apples and Morrisses mingling their blossoms, in the spring, with the green Forest oaks. As we reach the top, suddenly there opens out one long view. On the north-east rise the hills beyond Winchester; but the "White City" is hidden in their valley. To the east lies Southampton, with its houses by the water-side; and to the north, across the woods of Prior's Acre, gleam the green Wiltshire downs lit up by the sunlight.

Close to us, among its beeches, lies Castle Malwood, with its single trench and Forest lodge, where tradition and poets say Rufus feasted before his death; and down in the valley stands the Stone which marks the spot where he is said to have fallen.

It will be as well to repeat the story, as told by the two Chroniclers who give the fullest account, with all its omens and apparitions. The King had gone to bed on the evening of the 1st of August, and was suddenly awoke by a fearful vision. He dreamt that he was bled, and the stream of blood, pouring up to heaven, clouded the very day. His attendants, hearing his cries to the Virgin, rushed in with lights, and stayed with him all that night. Morning dawned: and Robert Fitz Hamon, his special friend, came to him with another dream, dreamt also that very night by a foreign monk then staying at the court, who had seen the King enter a church, and there seize the rood, tearing apart its legs and arms. For a time the image bore the insult, but suddenly struck the King. He fell, and flames and smoke issued from his mouth, putting out the light of the stars. The Red King's courage, however, had by this time returned. With a laugh, he cried, "He is a monk, and dreams for money like a monk: give him this," handing Fitz Hamon a hundred shillings. Still the two dreams had their effect, and "William hesitated to test their truth.[1] At dinner that day he drank more than usual. His spirits once more returned. He defied the dreams. In spite of their warnings, he determined to hunt. As he was preparing, his armourer approached with six brand-new arrows. Choosing out two, he cried, as he gave them to Walter Tiril, Lord of Poix and Pontoise, who had lately come from Normandy, "The best arrows to the best marksman." The small hunting-party, consisting of his brother Henry, William of Breteuil, Walter Tiril, and Fitz Hamon, and a few more, set out. As they are leaving the courtyard, a monk from St. Peter's Abbey at Gloucester arrives. He gives the King a letter from Serlo, the abbot. It told how a monk of that abbey had dreamt that he had seen the Saviour and all the host of heaven standing round the great white throne. Then, too, came the Virgin robed in light, and flung herself at the feet of her Son, and prayed Him, by his precious blood and agony on the cross, to take pity on the English; prayed too, as He was judge of all men, and avenger of all wickedness, to punish the King. The Saviour answered her, "You must be patient and wait: due retribution will in time befall the wicked." The King read it and laughed. "Does Serlo," he asked, "think that I believe the visions of every snoring monk? Does he take me for an Englishman, who puts faith in the dreams of every old woman?"[2] With this the party once more sets out into the Forest, the woods still green with all their deep summer foliage.

So they hunted all that noon and afternoon. The sun was now

setting. Tiril and the King were alone.[3] A stag bounded by: the King shot and slightly wounded the quarry. On, though, it still bounded in the full light of the setting sun. The King stood watching it, shading his eyes with his hands. At that moment another deer broke cover. Tiril this time shot, and the shaft lodged itself in the King's breast.[4] He fell without a word or groan, vainly trying to pull out the arrow, which broke short in his hand.

Thus perished William the Red. Tiril leapt on his horse. Henry galloped to Winchester, and the other nobles to their houses. One exception was there. William of Breteuil, following hard upon Henry to Winchester, honourably declared the rights of the absent Robert, to whom both Henry and himself had sworn fealty. William's body was brought on a cart to the cathedral, the blood from his wound reddening the road.[5] There the next morning[6] he was buried, unlamented, unknelled, and unaneled.[7]

So runs the story as told by the Chroniclers. And to this day popular tradition not only repeats their tale, but points to the places associated with the event. Below our feet lies the lonely glen of Canterton, where the King is said to have fallen.

Rufus's Stone.

The oak from which, as the legend runs, the arrow glanced, is long since dead, but a stone marks its site, now capped over with a hideous cast-iron case.[8] In the woods and in the village of Minestead still live some of the descendants of Purkess, who is reported to have carried the bleeding corpse in his charcoal-cart to Winchester along the road now known as the King's Road. Twelve miles away, on the extreme south-west boundary of the Forest, close to the Avon, stands a smithy, on the site of the one where, the legend says, Walter Tiril's horse was shod, and which, for that reason, to this day pays a yearly fine to the Crown: and the water close by, where the fugitive passed, is still called Tyrrel's Ford. And Rufus lies in Winchester Cathedral, his bones now mixed with those of Canute; and under a marble tomb, in the south aisle of the presbytery, sleeps his brother Richard, slain also like himself in the Forest.

So runs the story, unquestioned save here and there by some few faint doubts.[9] As to the tradition, I think we may at once set aside its testimony. The value of mere tradition in history weighs, or ought to weigh, nothing. Here and there tradition may be true in a very general sense, as when it says the Isle of Wight was once joined to Hampshire; but it is never particular in its dates, and is ever in too much hurry to compare facts. Tradition, as often as not, kills the murderer instead of the murdered; and makes the man who built the place to have been born there. Tradition is, in fact, the history of the vulgar, and the stumbling-block of the half-learned.

We will look at the broader bearings of the case. The first thing which strikes us is the fact that two other very near relatives of the Red King, his brother and his nephew, also lost their lives by so-called accidents in the New Forest. If we are to believe the Chroniclers, his brother Richard met his death whilst hunting there, according to one narrative, by a pestilential blast—surely, at the least, a very unsatisfactory account;[10] though, by another version, from the effects of a blow against a tree.[11] His nephew Richard was either wounded by an arrow through the neck, or caught by the boughs of a tree and strangled—a still more improbable death;[12] whilst, according to Florence of Worcester, he was killed by the arrow of one of his own knights.[13] We will only here pause to notice not only the extreme improbability, but the contradictory statements in both cases, which will not, of course, increase the value of the same evidence concerning Rufus.[14]

And now we will examine the version of his death. History is at all times subjective enough, but becomes far more so when written by unfriendly Chroniclers, who have good reasons for suppressing the truth. The story reads at the very first glance too much like a romance. In the first place, we have no less than three dreams, which are always effects rather than causes—after-thoughts rather than prophecies, well fitted to suit the superstition of the times, and to deceive the crowd. Then, too, we find the old device of the armourer craving the King to take six brand-new arrows, by one of which at the hand of his friend he is fated to fall on the very spot which his lather had laid waste, and where he is said to have destroyed a church.

It may of course be urged that all this is in accordance with what we know of the eternal power of the moral laws, that the sins of the fathers are ever visited upon the sons to the third and fourth generations, and that time ever completes the full circle of retribution. But the flaw is, that this special judgment is too special. "Divine vengeance" and "judgment of God," the Chroniclers cry out one after another, and this is thought sufficient to account for three so-called accidental deaths. The moral laws, however, never fall so directly as they are here represented. Their influence is more oblique. The lightning of justice does not immediately follow each peal of suffering.

Leaving, however, the Chroniclers' views to themselves, let us look further at some of the facts which peep out in the narrative. Why, in the first place, we naturally ask, if the King was shot by accident, did his friends and attendants desert him? Why was he brought home in a cart, drawn by a wretched jade, the blood, not even staunched, flowing from the wound, clotting the dust on the road? Why, too, the indecent haste of his funeral? Why, afterwards, was no inquiry as to his death made? Why, too, was Tiril's conduct not investigated? These questions are difficult to answer, except upon one supposition.

Let us note, also, that they are all ecclesiastics, to whom the revelations of the King's speedy end had been made known, and that their special favourite, Henry, succeeded to the throne in spite of his elder brother's right. It is, certainly, too, something more than singular that when the banished Anselm should visit Hugh, Abbot of Cluny, that the Abbot should tell him that during the past night he had seen William summoned before God and sentenced to damnation, and that the King's death immediately followed: that further, on the next day, when he went to Lyons, his chaplain should be twice told by a youth of the death of William before it took place.[15] More than singular, too, are those words of Fulchered, spoken so openly and so daringly, "The bow of God's vengeance is bent against the wicked; and the arrow swift to wound is already drawn out of the quiver."[16]

Either all these persons were prophets, or accessories to the murder, or—for there is one more solution—the Chroniclers invented this portion of the story. If we admit this last supposition, we cannot receive the other parts of the narrative without the greatest suspicion. We have almost a sufficient warrant to read them in an exactly opposite sense to what they were intended to bear.

Let us remember, also, that Flambard, Rufus's prime minister, who was universally hated by the clergy, and who had lately banished Godric, of Christchurch, into Normandy, was instantly stripped of his possessions by Henry, and Godric reinstated, and the banished Anselm recalled; and, lastly, and most important of all, that Tiril, who had just arrived from Normandy, was a friend of Anselm's,[17] and, further, that Alanus de Insulis, better known as le Docteur Universel, who lived not long after the event, actually says that in his opinion it was caused by treachery.[18] Surely all these facts and coincidences point but one way. All tend to show, as plainly as possible, that Rufus fell by no chance, but by a conspiracy of his prelates, who held the crozier in one, and the battle-axe in the other hand.[19] The cause of their hatred is at once supplied by his refusing to pay St. Peter's pence—denying the Pope's supremacy—banishing Anselm—promoting

Flambard—holding all the bishoprics and other offices which fell vacant[20]—by his cruelties to their different orders at Canterbury and Crowland, and throughout England, whose enmity died not with his death, but made them believe that the tower of Winchester Cathedral fell because they allowed him to be buried in its nave.

Reading, in the Chroniclers, the life of the Red King seems like rather reading a series of plots against it, not by the English, who were too thoroughly cowed to make the slightest resistance, but by his own prelates and barons.[21] His uncle Odo, Bishop of Bayeux, headed the first rebellion against him, as soon as he usurped the throne. William, Bishop of Durham, his own Minister, conspired against him. Bishop Gosfrith, with his nephew Robert, Earl of Northumberland, rebelled in the west. Roger Montgomery rose on the Welsh Marches. Roger Bigod in the eastern, and Hugo of Grentemesnil in the Midland Counties hoisted the flag of revolt.[22] Such was England at the beginning of his reign. In 1096, his own godfather, William de Aldrey, justly or unjustly, was accused of treason, and died on the gallows.[23] William, Count of Eu, kinsman to the King, suffered a worse fate for the same crime. His steward, William, also a kinsman of the King's, was hung on a rood. Eudes, Count of Champagne, forfeited his lands. Others not only shared the same fate, but were deprived of their eyesight.[24] His northern barons, headed by Robert of Mowbray, goaded to desperation by the Forest Laws, rose in revolt. Roger of Yvery, son of the Conqueror's favourite, led the Midland barons, and was obliged to fly, and all his vast estates, close to the New Forest, forfeited. Normandy, from whence Tiril had just come, swarmed with outlawed enemies, both churchmen and laymen. It was the nest where all the plots could be safely hatched.

Knowing all this, knowing, too, that the conspiracies became more frequent as his tyranny increased, we can scarcely avoid coming to but one conclusion as to his death.

It might suit the policy of the times to throw the guilt on Tiril, but Tiril certainly did not shoot the arrow. We have his own most solemn declaration to various people, and especially, not once but often, to Suger, the well-known Abbot of St. Denis, when he had nothing to gain or lose, that he had on the day of the King's death not only not entered that part of the Forest, but had not so much as even seen him.[25]

Tiril, however, was certainly implicated in the plot. His haste to leave the country arose, probably, not so much from a wish to escape as to convey the news of the success to Normandy: and popular tradition mistaking the cause, with its usual inaccuracy, fixed on the wrong person as the assassin. In after years, however, from some scruple of conscience, he expiated his share in the murder by a pilgrimage to the Holy Land.

Who shot the fatal arrow we know not, and, perhaps, shall never know. We must not expect to get truth in history,—only, at the best, some faint glimmering. All is here confusion and darkness. John of Salisbury, who lived about the middle of the twelfth century, says it was as little known who killed the King as who slew Julian the Apostate.[26] The very spot where he fell is doubtful. One thing, however, seems certain, that he was slain, not, as the Chroniclers say, because his father made the New Forest, but through his own cruelties and excesses, by which he outraged both friend and foe.

It is not single passages which alone leave this impression, but still more the cumulative force of the evidence. The fact that all were gainers by his death, and the general abhorrence of the tyrant, are in themselves strong reasons. Not one, but all parties were bound together against him by the strongest of covenants—hatred. The marked and bitter prophecies, which would not have been uttered were not their fulfilment ensured,—the suspicious silence on all important points,—the pretended dreams and omens,—the abandonment of the body,—the want of any inquiry into the cause of death,—the connection between the Church party and Anselm with Henry I., and Anselm's connection again with Tiril, all serve to show the depth and darkness of the plot.

His life throws the best light on his death. Read by it, by the extortions and atrocities which he committed, by the universal hatred in which he was held, the conclusion is inevitable. Years of violence were the prelude to a violent end. The many failures in open revolt seem only to have taught the lesson of greater caution. And treachery at last succeeded, where plain courage had so often failed.

Direct proof of the murder cannot be had, and must not be expected. Every one was interested in keeping that a secret by which all alike profited. To have declared it, would have covered the Crown with disgrace, and stained the hands of the Church.

Their own absurdities and contradictions form the best refutation of the common accounts. In details they are irreconcilable with each other. According to one, the King was alone with Tiril; to another, with all his attendants. One narrative declares that the arrow glanced from a boar, a second from a stag, a third from a tree. Even if we accept them, then either the power of prophecy lasted much longer than is commonly supposed, or, as we have said, the clergy were accessories to the murder;—we have no other choice. The last of these solutions is fatal to the common belief; and very few persons would, I suppose, venture upon the first. Nevertheless, the monk of Gloucester's dream was not yet to be fulfilled. The hour was not yet at hand for England's deliverance. As the Parliamentary party said in Charles the First's time—"Things must become worse before they can mend." England had, therefore, to undergo a tyranny for more than a century longer—till the evil became its own cure. Good was at length accomplished. Out of all the woe and wretchedness came the Bill of Rights and the Charta de Forestâ.

View from Castle Malwood.

Footnotes

1. William of Malmesbury: *Gesta Regum Anglorum*. Ed. Hardy, tom, ii., lib. iv., sect. 333, p. 508.
2. Vitalis: *Historia Eccl.*, pars, iii., lib x., cap. xii., in Migne: *Patrologicæ Cursus*, tom, clxxxviii. pp. 751, 752; where occurs (pp. 750, 751) a most remarkable sermon, on the wrongs and woes of England, preached at St. Peter's Abbey, Shrewsbury, on St. Peter's Day, by Fulchered, first abbot of Shrewsbury, a man evidently of high purpose, ending with these ominous words:—" The bow of

God's vengeance is bent against the wicked. The arrow, swift to wound, is already drawn out of the quiver. Soon will the blow be struck but the man who is wise to amend will avoid it." Surely this is more than a general denunciation. On the very next day William the Red falls.
3. Malmesbury, as before quoted, p. 509. Vitalis, however, in Migne, as before, p. 751, says there were some others.
4. William of Malmesbury says nothing about the tree, from which nearly all modern historians represent the arrow as glancing. Vitalis, as before, p. 751, expressly states that it rebounded from the back of a beast of chase (*fera*), apparently, by the mention of bristles (*setæ*), a wild-boar. Matthew Paris (Ed. Wats., tom. i. p. 54) first mentions the tree, but his narrative is doubtful.
5. Malmesbury, as before, p. 509. The additions that it was a charcoal-cart, as also the owner's name, are merely traditional.
6. The *Chronicle*. Ed. Thorpe, vol. i. p. 364.
7. Vitalis, as before, p. 752. Neither William of Malmesbury nor Vitalis, who go into details, mentions the spot where the King was killed. The *Chronicle* and Florence of Worcester most briefly relate the accident, though Florence adds that William fell where his father had destroyed a chapel. (Ed. Thorpe, vol. ii. p. 45). Henry of Huntingdon (*Historiarum*, lib. vii., in Saville's *Scriptores Rerum Anglicarum*, p. 378) says but little more, dwelling only on the King's wickedness and the supernatural appearance of blood. Matthew Paris brings a bishop on the scene, as explaining another dream of the King's, and gives the King's speech of "trahe arcum, diabole" to Tiril, which has a certain mad humour about it, as also the incident of the tree, and the apparition of a goat (*Hist. Major. Angl.* Ed. Wats., pp. 53, 54), which are not to be found in Roger of Wendover (*Flores Hist.* Ed. Coxe, tom. ii., pp. 157-59), and therefore open to the strongest suspicion. Matthew of Westminster (*Flores Hist.* Ed. 1601, p. 235) follows, in

most of his details, William of Malmesbury. Simon of Durham (*De Gestis Regum Anglorum*, in Twysden's *Historia Anglicanæ Scriptores Decem*, p. 225), as, too, Walter de Hemingburgh (Ed. Hamilton, vol. i. p. 33), and Roger Hoveden (*Annalium Pars Prior*, in Saville's *Rerum Anglicarum Scriptores*, pp. 467, 468), copy Florence of Worcester. So, too, in various ways, with all the later writers, who had access to no new sources of information. Peter Blois, however, in his continuation of *Ingulph* (Gales's *Rerum Anglicarum Scriptores*, tom. i. pp. 110, 111; Oxford 1684) is more vivid, and adds that the dogs were chasing the stags up a hill; but his whole book is very doubtful, and his account in this particular instance is irreconcilable with the others. They all, however, with the exception of the *Chronicle* and William of Newburgh (Ed. Hamilton, vol. i. book i., ch. ii., p. 17), who are silent, agree in saying that Tiril committed the deed by accident. Of later writers, Leland, in his *Itinerary* (vol. vi. f. 100, p. 88) states that the King fell at Thorougham, where in his time there was still a chapel standing, evidently meaning Fritham, called Truham in *Domesday*. Gilpin (*Forest Scenery*, vol. i. p, 166) mentions a similar tradition; so that there is a very reasonable doubt as to the spot itself being where the Stone stands, especially since, with the exception of the vague remark of Florence, none of the best Chroniclers say one word about the place. Thierry, in many minor particulars, follows Knyghton, whose authority is of little value, and I have therefore omitted all reference to him.

8. Very much against my inclination, I give a sketch of the iron case of the Stone, which the artist has certainly succeeded in making as beautiful as it is possible to do. The public would not, I know, think the book complete without it. It stands, however, rather as a monument of the habit of that English public, who imagine that their eyes are at their fingers' ends, and of a taste which is on a par with that of the designer of the post-office pillar-boxes, than of the Red King's death; for the spot where he fell

is, as we have seen from the previous note, by no means certain. We must, too, remember that there is no mention made by the Chroniclers of Castle Malwood, but the context in Vitalis, as also the late hour mentioned by Malmesbury when William went out to hunt, show that he was at the time staying somewhere in the Forest.

9. See, as before, Lappenberg's *History of England under the Norman Kings*, pp. 266-8; and Sharon Turner's *History of England during the Middle Ages*, vol. iv. pp. 166-8.

10. "Tabidi aëris nebulâ" are the words of William of Malmesbury. (*Gesta Regum Anglorum*. Ed. Hardy, tom, ii., lib. iii., sect. 275, pp. 454, 455.)

11. *Gul. Gemeticensis de Ducibus Normannorum*, lib. vii., cap. ix. To be found in Camden's *Anglica Scripta*, p. 674.

12. This seems to be the meaning of a not very clear passage in William of Malmesbury. Same edition as before, p. 455. Vitalis, however, *Historia Ecclesiastica*, pars 3, lib. x., cap. xi. (in *Migne, Patrologiæ Cursus Completus*, tom. clxxxviii. pp. 748, 749), says he was shot by a knight, who expiated the deed by retiring to a monastery, and speaks in high terms both of him and his brother William, who fell in one of the Crusades.

13. Ed. Thorpe, vol. ii. p. 45. Lewis, in his *Topographical Remarks on the New Forest*, pp. 57-62, is hopelessly wrong with regard to Richard, the son of Robert, a grandson of the Conqueror, whom he calls Henry, and confounds at p. 62 with his uncle; and makes both William of Malmesbury and Baker (see his *Chronicle*, p. 37, Ed. 1730) say quite the reverse of what they write.

14. As I am not writing a History of England during this period, my space will not permit me to enter into those details which, when viewed collectively, carry so much weight in an argument, but at all events, it will be well for some of my readers to bear in mind the character of William II., who in a recent work has lately been elevated into a hero. Without any of his father's

ability or power of statesmanship, he inherited all his vices, which he so improved that they became rather his own. From having no occupation for his mind, he sank more and more into licentiousness and lust. ("Omni se immunditia deturpabat," is the strong expression of John of Salisbury. *Life of Anselm*, part ii. ch. vii., in Wharton's *Anglia Sacra*, tom. ii. p. 163. See, also, Suger, *Vita Lud. Grossi Regis*, cap. i., in Bouquet, *Recueil des Historiens des Gaules et de la France*, tom. xii. p. 12. D. E.) Being lustful, he naturally became cruel; not as his father was, on, at least, the plea of necessity, but that he might enjoy a cultivated pleasure in gloating over the sufferings of others. From being cruel, too, he became, in its worst sense, an infidel; not from any pious scruple or deep conviction, but simply that he might indulge his passions. (See that fearful story of the trial of forty Englishmen told in Eadmer: *Hist. Nov.*, lib. ii., p. 48, Ed. 1633, which illustrates in a twofold manner both his cruelty and his atheism.)

To a total want of eloquence he joined the most inveterate habit of stammering, so that, when angry, he could barely speak. His physical appearance, too, well harmonized with his moral and mental deformities. His description reads rather like that of a fiend than of a man. Possessing enormous strength, he was small, thick-set, and ill-shaped, having a large stomach. His face was redder than his hair, and his eyes of two different colours. His vices were, in fact, branded on his face. (Malmesbury, Ed. Hardy, tom. ii., lib. iv., sect. 321, p. 504, whom I have literally translated.)

Let us look, too, at the events of his reign. Crime after crime crowds upon us. His first act was to imprison those whom his father had set free. He loaded the Forest haws with fresh horrors. Impartial in his cruelty, he plundered both castle and monastery (*The Chronicle*, Ed. Thorpe, vol. i. p. 364). He burnt out the eyes of the inhabitants of Canterbury, who had taken the part of the monks of St Augustin's. At the very mention of his approach the people fled (Eadmer: *Hist. Nov.*, lib. iv. p. 94). Unable himself

to be everywhere, his favourites, Robert d'Ouilly harried the middle, and Odineau d'Omfreville the north of England; whilst his Minister, Ralph Flambard, committed such excesses that the people prayed for death as their only deliverance (*Annal. Eccles.* Winton., in Wharton's *Anglia Sacra*, tom. i. p. 295).
As *The Chronicle* impressively says, "In his days all right fell, and all wrong in the sight of God and of the world rose." Norman and English, friend and foe, priest and layman, were united by one common bond of hatred against the tyrant. It could only be expected that as his life was, so his death would be; that he would be betrayed by his companions, and in his utmost need deserted by his friends.
15. Eadmer: *Vita Anselmi*, Ed. Paris, 1721, p. 23. John of Salisbury: *Vita Anselmi*, cap. xi.; in Wharton's *Anglia Sacra*, tom. ii. p. 169. William of Malmesbury: Ed. Hardy, vol. ii., b. iv , sect. 332, p. 507.; and Roger of Wendover, Ed. Coxe, vol. ii. pp. 159, 160.
16. Vitalis: *Historia Ecclesiastica*, pars 3, lib. x.; in Migne, *Patrologiæ Cursus Completus*, tom, clxxxviii., pp. 750 D, 751 A. See previously, p. 94, foot-note.
17. Eadmer: *Vita Anselmi*, Ed. Paris, 1721, p. 6.
18. Baxter, in his Preface to his *Glossurium Antiquitatum Britannicarum*, Ed. 1719, p. 12, entirely misquotes Alanus de Insulis (see *Prophetica Anglicana Merlini Ambrosii cum septem libris explanationum Alani de Insulis.* Frankfort, 1603. Lib. ii. pp. 68, 69), and completely misunderstands the passage. Alanus, however (p. 69), seems to have no doubt that the King fell by treachery,—"spiculo invidiæ," as was foretold by Merlin, though he gives no other reason; and which by itself, resting on nothing further, would carry no weight. His account, though, of the general detestation of the Red King immediately before his death, as also the conversation of Hugh, Abbot of Cluny, with Anselm (p. 68), is very suggestive, especially by the way in which it is introduced Alanus must have possessed far too shrewd an

intellect to have believed in Merlin; though it might have suited his purpose to have appeared to have so done, as a veil and a blind, so that he might better say what his high position and authority would not in any other form have well permitted, but which still give to many points, as here, enormous significance and weight. The first person besides Alanus who seems to hint at treachery is Nicander Nucius (*Second Book of Travels*, published by the Camden Society, pp. 34, 35), but his account is too vague to be of any service. We should, however, constantly bear in mind, with Lappenberg, that the best authority. *The Chronicle*, simply relates that the King was shot at the chase by one of his friends, without any allusion to an accident. Not one word or fact else is given, except the appearance of a pool of blood in Berkshire (at Finchhamstead, according to William of Malmesbury), which we know, from other sources, was supposed to foretell some calamity, and which phenomenon science now resolves into merely some species of *alga*, probably either *Palmella cruenta* or *Hæmatococcus sanguineus*. Eadmer, with some others, in his *Historia Novorum*, lib. ii. (Migne: *Patrolugiæ Cursus Completus*, tom. clix. p. 422 B) mentions a report, prevalent at the time, that the King accidentally stumbled on an arrow. Then follows, in the very next book (Migne, as before, p. 423 B), a singular passage, to be found also in his *Life of Anselm*, book ii. ch. vi. (Migne, as before, tom. clviii. p. 108 D), where, on the news of the Red King's death, Anselm bursts into tears, and, with sobs, cries, "Quod si hoc efficere posset, multo magis eligeret se ipsum corpore, quam illud, sicut erat, mortuum esse." Whether this wish sprang from the effects of some pangs of conscience as to William's death, or from an honourable feeling of natural emotion under the circumstances, as suggested by Sharon Turner, it is hard to determine. From John of Salisbury (*Vita Anselmi*, pars ii., cap. xi,, in Wharton's *Anglia Sacra*, tom. ii. p. 169), it would seem that Anselm thought that he was the direct cause, through God, of his death. Wace, quoted by Sharon Turner (vol. iv. p. 169), says that a woman prophesied to Henry his speedy accession to the throne; but I am not inclined to put any faith in this story, especially as Wace's account is in poetry, where a prophetical speech might after the event be given dramatically true, without being so historically. Gaimar (*MS. Bib. Reg.* 13, A 21, also quoted by Turner), a rhymer, nearly contemporary, sings, "that the other archers said that the shaft came from Walter Tiril's bow." No one, however, was likely to declare, for so many reasons, that the King was murdered. We must not expect such a statement, or even look for it in the Chroniclers; we must seek for it in the contradictions, and absurdities, and prophecies which have gathered round the event.

19. Let no one be startled at the fact of ecclesiastics being assassins. We have on record during this very reign the deliberate confessions by monks of plots to murder their abbots, deeming they were doing God a service. We must further keep steadily in mind that prelates then united in their own persons both sacred and military offices. How much Henry was under the influence of the monasteries his marriage and his various appointments show. Their power was enormous. In fact, I believe that the Conqueror owed his success as much to them as Rufus his death, and Henry his crown.

20. At the time of his death he held in his hand the archbishopric of Canterbury, the bishoprics of Winchester and Salisbury, besides eleven abbacies, all let out to rent. *The Chronicle*, Ed. Thorpe, vol. i. p. 364.

21. The Chronicle, Ed. Thorpe, vol. i. p. 356.

22. William of Malmesbury, Ed. Hardy, tom, ii., lib. iv., sect. 306, p. 488.

23. The same, tom., ii., lib. iv., sect. 319, p. 502.

24. The Chronicle, Ed. Thorpe, vol. i. p. 362.

25. Suger: *Vita Lud. Grossi Regis*, cap. i (to be found, as before,

in Bouquet, tom. xii. p. 12 E.) See, also, John of Salisbury: *Vita Anselmi*; Migne: *Patrologiæ Cursus Completus*, tom, cxcix., cap. xii., p. 1031 B.; or, as before, in Wharton's *Anglia Sacra*, tom. ii. p. 170.

26. Quoted by Sharon Turner: *History of England*, vol. iv. p. 167. See, as before, Migne: tom, cxcix., cap. xii., p. 1031 B.

CHAPTER X.

THE NORTHERN PART.—STONEY-CROSS, BRAMBLE HILL, FRITHAM, BENTLEY, EYEWORTH, SLODEN.

View in Studley Wood

If any one wishes to know the beauty of the Forest in autumn, let him see the view from the high ridge at Stoney-Cross. Here the air blows off the Wiltshire Downs finer and keener than anywhere else. Here, on all sides, stretch woods and moors. Here, in the latter end of August, the three heathers, one after another, cover every plain and holt with their crimson glory, mixed with the flashes of the dwarf furze. And a little later the maples are dyed, yellow and russet, by the autumn rains, and the beeches are scorched to a fiery red with the first frost, and the oaks renew, but deeper and more gloriously, the golden lights of spring, till the great woods of Prior's Acre and Daneshill burn with colour; every gleam of sunshine, and every passing shadow, touching them with fresher and stranger beauty.[1]

To the east, about two miles along the Southampton Road, lies the village of Cadenham, famous for its oak, which, like the Glastonbury thorn, buds on Christmas Eve. The popular tradition in the neighbourhood runs, that, as the weather is harder, it shows more leaves, and, refusing the present chronology, only buds on Old Christmas night. As in most things, there is some little truth in the story. Doubtless, in some of the mild winters which visit Hampshire, the tree shows a few buds, as at that time I have seen others do in various parts of the Forest. Of course, they are all nipped by the first approach of severe weather, which, however, seldom happens on the warm south-west coast till the new year.

Down in the valley to the left of Rufus's Stone rise the woods of the Long Beeches, and Prior's Acre, and Daneshill or Dean's Hell, where the word Hell (from *helan*, to cover) means nothing more than the dark place, like the Hellbecks in Yorkshire.[2] Beaten paths and walks stretch into the woods in every direction. Perhaps one of the prettiest is over Coalmeer Brook, and then through the thick beeches of Coalmeer Wood, where the honey buzzard builds, till we come to the King's Gairn stream, where the Bracklesham Clays, teeming with fossils, may, by digging, be reached.[3]

Brook Common now opens before us. At its farther end stands Brook Wood, with its fine hollies and durmast oaks (*Quercus sessiliflora*). Passing the High Beeches to our left, we reach Shepherd's Gutter, a small stream, where the Bracklesham beds again crop out with their blue and slate-coloured clays.

Going on through more woods, and then by clumps of old hollies and yews, we come to Bramble Hill. Perhaps, just above the Lodge, on the top of the hill, we gain the most extensive view of the Forest. Before us spreads one vast sea of woods, broken in the front by Malwood Ridge, and Brochis Hill, and then rolling its flood of green over Minestead Valley, and rising again wave-like, at Whitley, till lost among the moors, whilst the Isle of Wight hills seam the blue sky with their dark outlines.

The village of Bramshaw, just a little way beyond, stands partly in both Hampshire and Wiltshire, and forms the Forest boundary. From its woods in former times the shingles for roofing Salisbury Cathedral were cut. Its church, although prettily situated, is scarcely worth seeing. Only an Early-English window at the east end, and an arch on the south side, remain of the old building, now defaced by every variety of modern ugliness. In the churchyard stands a fine yew; and a buttress on the north side is completely covered with the lovely common spleenwort.

Coming back, however, to Stoney-Cross, we will now go westward. Stoney-Cross itself consists of but a few tumbledown cottages, inhabited principally by the Forest workmen. Just beyond the last of them let us stop for a moment. To the south stretch more woods—Stonehard, with its views across the valley, to the oaks of Wick and the plain of Acres Down, looking over Rhinefield and the valley of the Osmanby Ford, beyond Wootton, to the Needle Rocks, mass upon mass of

View in Puckpits

woods. To the right of it lies Puckpits, where the badger breeds, and the raven used to build, and where still on a summer morning the honey buzzard comes flying up from Mark Ash, and, circling for hours round the trees, will again fly back to its favourite haunt.

All these woods there are for rambles, flushed in the spring with wood-anemones and wood-sorrel, set in the green moss and the greener heather of the bilberry. Nowhere, too, in the Forest, than in these woods, have I seen more lovely sunsets. Through some deep-cut oriel of the trees have I watched the sun begin to sink, each moment burning brighter, and then suddenly its great brand of fire would fall, reddening each tree trunk, and crimson billows of clouds come rolling eastward.

Instead of following the Ringwood Road, beautiful as that is in many parts, especially at Woody Bratley, with its old thorn trees, we will turn off to the right. To the west now rises Ocknell Wood, and its clump of firs, a well-known landmark, and beyond that lies the new Slufter Inclosure, and Bratley Plain, with its great graveyard of barrows. In front of us stretches the East Fritham Plain, with its three barrows, locally called "butts," the central known as Reachmore. At the second mound we will go into North Bentley Wood, following the wood-cutter's track. Very wild and unfrequented is this. Here a stray deer will bound across the road; and sometimes a small herd of as many as six or seven are browsing on the ivy clinging to some tree just felled, startled at the slightest sound, and trooping off down the glades. The grey hen rises up at our feet from the heather; and, as we enter the wood, the woodpecker shrieks out his shrill laugh, whilst a buzzard is heavily sailing over the trees.

The road winds on through the valley amongst oaks flecked with silver flakes of moss, broken here and there by open glades and green spaces of fern. At last, we reach Queen's North Lawn, which leads us on the right to Fritham, standing on the hill top. In the valley below lies Eyeworth Lodge, with the powder mills lately built; the Ivare of *Domesday*, and still so called by the peasantry, afterwards Yvez, where Roger Beteston, in the reign of Henry III., held some land by the service of finding litter for the King's bed and hay for his horse whenever he came here to hunt.[4]

Fritham is thoroughly in the Forest; and few spots can equal it in interest. It may be the very place where Rufus fell:[5] but whether or no, close round it lie the barrows of the Kelt, and the potteries of the Roman, covering acres of ground, at Island's Thorn and Crockle, and Sloden and Black Bar, with the banks which mark the sites of the workmen's houses.[6] Close round it, too, encircling it on all sides, rise the woods of Studley, with their great beeches, and Eyeworth, famous for its well. Going along the West Fritham Plain we come to Sloden, with its thick wood of yews, standing, massive and black, in all their depth of foliage, mixed, in loveliest contrast, with clumps of whitebeams. Below runs the brook, flowing under Amberwood, and winding among dark groups of hollies, lost at last in the deep gorge, shut in by the hills of Goreley and Charlford.

The best way to reach Fordingbridge is either to go by Ashley Lodge, and so through Pitt's Wood, and between the high, bare, half mountainous hills of Chilly and Blissford, coming out upon the turnpike-road near Blissford Gate; or to follow the side of the Amberwood stream towards some scattered houses, called Ogdens.

Here we leave the Forest, and its moors and woods, and, mounting Goreley Hill, see below us the church of Fordingbridge, and the Avon winding among its meadows. To the south Hengistbury Head lifts itself up in the distant horizon; and beyond it again, but more to the west, stretches the blue line of the Portland Hills. To the north swell the rounded forms of the Wiltshire downs, and the spire of Salisbury starts out from the midst, and behind it towers the crag of Old Sarum.

Yews and Whitebeams in Sloden.

Footnotes

1. The best way to know this part of the Forest, is to go to Stoney-Cross itself, and stay at the inn, once a well-known house in the old posting days, and lately fitted up for visitors.
2. The word, however, is going out of use, and is more generally now softened into hill. We meet with it in the perambulation of the Forest made in the twenty-second year of Charles II.—"The same hedge reaches Barnfarn from the right hand, right by Helclose, as far as to a certain corner called Hell Corner."
3. For the geology of this part of the Forest see chapter xx.
4. *Testa de Nevill*, p. 237 b. 130. See, also, p. 235 b. (118). As in

other parts of England, throughout the Forest, as we have seen at Lyndhurst and Brockenhurst, were similar feudal tenures. Some held their lands, as the heirs of Cobbe, at Eling, by finding 50; and others, again, as Richard de Baudet, at Redbridge, 100 arrows. Testa de Nevill, as in the first reference; and p. 238 a. (132.)

5. See previous chapter, p. 96, foot-note.

6. For some account of the contents of these barrows and potteries, see chapters xvii. and xviii.

CHAPTER XI.

THE VALLEY OF THE AVON.—FORDINGBRIDGE, CHARFORD, BREAMORE. IBBESLEY, ELLINGHAM, RINGWOOD, SOPLEY.

The Valley of the Avon from Castle Hill

THE Valley of the Avon should certainly be seen, both because large parts of its manors and villages once stood in the Forest, as also for the contrast which it now affords to the neighbouring Forest scenery. Nothing can be so different to the moors we have just left as the Valley. Though close to them, you might imagine you were suddenly transported into one of the the Midland

Counties, and were walking by the side of the Warwickshire, instead of the Wiltshire Avon. In the place of wild heathery commons and furzy holts, deep lanes wind along by comfortable homesteads, thatched with Dorsetshire reed. Instead, too, of dark oak and beech woods, thick hedges are white in the spring with the scattered spray of the blackthorn, and orchards glow with their crimson wreaths of flowers.

Fordingbridge, formerly nothing else but Forde, now known to all fishermen for its pike and trout, in former days held the high-road into the Forest. On the bridge the lord of the manor, during the fence months, was obliged to mount guard, and stop all suspected persons, who could only on the north-west leave the Forest this way.[1]

In *Domesday* it possessed a church and two mills, assessed at 14*s.* 2*d.* Though all its beech and oak woods, worth, on account of the pannage for swine, 20*s.* a year, were afforested, only three virgates of land were taken. Yet, notwithstanding this loss, it still paid the same rental as in Edward the Confessor's reign.

The old hospital, dedicated to St. John, was dissolved by Henry VI., and its revenues annexed to St. Cross, near Winchester.[2] The church stands on the extreme south-west side of the town, with its avenue of limes, and its yews, now spoilt by being clipt. The windows of the nave are Early Decorated, whilst those of the clerestory are Perpendicular. Against the north pillar of the south chancel arch is fixed a late brass. The upper part of the east window is spoilt by its ugly Tudor headings, and the lower portion by the Commandment tables. The high-pitched open Perpendicular roof of the north chancel, however, possesses some real interest, both on account of its height and its richness of detail,—the tie-beams faced with mouldings, and the spaces above ornamented with tracery, and the braces below also carved, and the purlins enriched with bosses, whilst carved projecting figures bear up the whole.

Before, however, the traveller leaves Fordingbridge he should go to Sandyballs and Castle Hill, where are still the remains of a camp, and traces of habitations, probably used in turn by Kelts, Romans, and West-Saxons, and where, perhaps, Ambrosius entrenched himself before the battle of Charford. From here is one of the best views of the Valley. Behind us stands Godshill inclosure, and the Forest with its dark moors and woods. Below winds the Avon, with its orchards nestling on the hill side, stretching its silver coil of waters along the green meadows, the sunlight gleaming on each bend and turn.

Looking up the stream, the village of Wood Green, and the woods of Hale, and the two Charfords, one by one appear. Charford is especially noticeable, formerly Cerdeford, without doubt the Cerdices-ford of *The Chronicle* and of Florence. Here it was for the last time that the gallant Ambrosius Aurelianus, Prince Natan-Leod, father of the great Arthur of Mediæval legends, after his many defeats, rallied the forlorn hope of the Romanized Kelts. Here, too, he fell on the greensward by the side of the Avon, with five thousand of his men, and was buried at Amesbury, which still preserves his name. Of the battle we know nothing—know only this, that the Keltic power in Wessex was broken, and that from henceforth the land from Winchester to Charford was called Natan-lea.[3]

Close to Charford lies Breamore,[4]—the last of the Forest villages to the north-west mentioned in *Domesday*—with the ruins of its fine Elizabethan hall, burnt down only a few years since, and its church standing in a graveyard full of old yews and laurels. The church has been most shamefully disfigured—stuccoed outside, and whitewashed within. Still it is worth seeing. A Norman doorway, another proof that the Conqueror did not destroy every church in the district, stands inside the south porch. A piscina, and brackets for images, still remain in the chancel.

Returning to Fordingbridge we pass through Burgate, formerly belonging to Beaulieu Abbey, where the dogs of the Lord of the Manor, like those of the Abbot of the Monastery, were allowed to go "unlawed." The base of the old village cross still remains, but the head was, not long ago, broken to pieces to mend the roads.

Our way from Fordingbridge lies by the side of the Avon, with the new chapel of Hyde or Hungerford standing on the top of the Forest range of hills. The road soon brings us to Ibbesley, the prettiest of villages in the Valley, with its cottages by the road-side, and their gardens of roses and poppies and sweet peas, and their porches thatched with honeysuckle. Three great elms overhang the river, spanned by the single arch of its bridge; whilst the stream pours sparkling and foaming over the weir into the water-meadows, and in the distance the tower of Harbridge rises out from its trees.

The sketch which is given at the end of this chapter is taken lower down in the fields, and shows another view not so well known. But the whole river is here full of beauty, winding, scarce knowing where, among the flat meadows, one stream flowing one way, and one another, and then all suddenly uniting, coming up with their joined force against the steep banks, dark in the shade of the trees; and, being repulsed, flowing away again into the meadows, white with flocks of swans, and fenced in by green hedges of rushes and yellow flags.

Going on we reach the avenue of elms which brings us to the Ellingham cross roads. Turning up the lane to the left we presently come to Moyles Court, just on the boundary of the Forest, looking out upon the woods of Newlyns and Chartley. Here lived Alice Lisle, and here are shown the hiding-places where, after the battle of Sedgemoor, she concealed Hicks and Nelthorpe. The house is sadly out of repair; the oak floors, and part of the fine old staircase, and the wainscoting of many of the rooms have been taken away; the old tapestry is destroyed and the iron gates rusted and broken. Still the private chapel remains, with its panelling and carved string-course of heads, and its "Ecce Homo" over the place where the altar once stood.[5]

The story of Alice Lisle needs no telling. She was found guilty of high treason not by the jury, but by the judge,—the infamous Jeffreys,—and was condemned, for an act of Christian kindness, to worse than a felon's death.

In Ellingham churchyard, close to the south porch, stands a plain brick tomb under which she, and her daughter Anne Hartell, lie, with the simple words, "Alicia Lisle dyed the second of September, 1685;" and round the tomb, weaving its ever green chaplet, grows the little rue-leaved spleenwort.

But a nobler monument has been raised to her in our Houses of Parliament. In the Commons' corridor she stands, bent with age, resting on her staff, with a gentle placidness shining in her face, unmoved by any fears for the future, but caring only to do what her heart feels to be right; whilst on the opposite wall, painted by the same hand, lives another of those Englishwomen of whom we may be proud,—Jane Lane, who, in her loyalty, would as willingly have sacrificed herself for one of the most ungrateful of princes, as Alice Lisle for the poor Puritans. And about eight miles away, across the Avon, in Dorsetshire, between two fields on Woodlands Farm, runs an old-fashioned double hedge, the central ditch choked up with hazel, and holly, and the common brake. About midway down, half in the ditch and half in the hedge, stands a pollarded ash, now bored into holes by the woodpeckers. This is Monmouth's Ash, and close to it, in the ditch, the duke, the miserable cause of so much misery, was seized, hid among the fern and brambles.[6]

To the ecclesiologist the little church of Ellingham (Adeling's hamlet) is full of interest. Within stands the old covered carved pew of Moyles Court, and a monument to one of its former owners. The plain rood-screen, with the stand for the hour-glass, and the marks of the pulpit still remain, formerly, as we can still see, painted blue like the chancel. On the south wall traces of the staircase to the rood-loft, as well as the entrance from the outside, are also still visible. In the chancel the Early-English windows have been sadly mutilated. Over the communion-table hangs a picture of the Day of Judgment, plundered from some church in Port St. Mary, in the Bay of Cadiz, whose bad execution is only exceeded by its indecent materialism. In the south chancel wall is a double piscina. On the walls above the rood-screen, the twenty-first verse of the twenty-fourth chapter of Proverbs, and the twenty-fourth verse of the third chapter of Galatians, according to the version of the Geneva Bible, are roughly painted.[7]

As in all the other churches of the district, the churchwardens have here from time to time shown their natural attachment to ugliness. The Early-English triplet at the east end has been blocked up, the gravestones in the chancel defaced, and a brick porch patched on at the south side.

The road now winds on by low water-meadows, pastured by herds of cattle, past Blashford Green, till we reach Ringwood, the Rinwede of *Domesday*.[8] Here, at the Grammar School, was Stillingfleet educated. Here Monmouth wrote his three craven letters to James, the Queen Dowager, and the Lord Treasurer, imploring them to save that life which it was a disgrace to own.

The old church has been pulled down, and a new one, modelled in every particular after it, has been built on its site. A church ought doubtless to tell its own date by its style. Yet it is far better that we should copy a moderately good specimen than increase the number of modern abortions. At all events, this is faithfully restored, though utterly spoilt by the heavy galleries which flank it on every side. The Early-English chancel, with its recessed arcade, springing from polished shafts of black Purbeck marble, well shows the beauty of the original design; whilst, on the chancel floor, lies a fine brass of the fifteenth century to John Prophete, which, however, has been most shamefully defaced. The body is robed in a cope broidered with figures of saints—St. Michael, and the Virgin and Child, St. Peter and St. Paul, St. Catharine and St. Faith, St. George and "Sancta Wefreda." The head, with the hood thrown back, rests on a cushion, whilst the cope is clasped with a morse, enriched with an effigy of the Saviour, crowned with a halo of light.

The Avon at Ibbesley.

Footnotes

1. Lewis: *Topographical Remarks on the New Forest*, p. 80, foot-note. I have not, however, been able to find his authority. A tradition of the sort lingers in the neighbourhood. Blount (*Fragmenta Antiquitatis*, Ed. Beck with, p. 115. 1815) says that Richard Carevile held here six librates a year of land in chief of Edward I., by finding a sergeant-at-arms for forty days every year

in the King's army. See, also, the *Testa de Nevill*, p. 231 (101), No. 3.

2. Dugdale: *Monasticon Anglicanum*, Ed. 1830, vol. vi., part, ii., p. 761. Leland, however (*Itin.*, vol. iii., f. 72, p. 88, Ed. Hearne), says it was given to King's College, Cambridge.

3. *The Chronicle*, Ed. Thorpe, vol. i. p. 26. *Florence of Worcester*, Ed. Thorpe, vol. i. p. 4.

4. This is the Brumore of *Domesday* (same edition as before, p. iv. a), and then belonged to the manor of Rockbourne, and was held by the Conqueror, as it had also been by Edward the Confessor. Two hydes and a half, and a wood capable of supporting fifty swine, were taken into the Forest. From the mention of a priest (*presbyter*), who received twenty shillings from some land in the Isle of Wight, there would seem to have been a church, in all probability situated, as the old yew would show, in the present churchyard, and of which the Norman doorway may be the last remains.

The Valley of the Avon, as was mentioned in chapter v., p. 51, footnote, appears from its nature to have been, with the exception of the east coast, the most flourishing district of any in the neighbourhood of the Forest. It is worth, however, noticing that many of its mills were rated not only by a money value, but by the additional payment of so many eels. Thus at Charford (Cerdeford) the mill is assessed at 15*s*. and 1,250 eels, and at Burgate (Borgate) the mill paid 10*s*. and 1,000 eels, whilst at Ibbesley (Tibeslei) the assessment was only 10*s*. and 700 eels (*Domesday*, as before, pp. xix. a, iv. b, xviii. a). The latter place had two hydes, and Burgate its woods and pasture, which maintained forty hogs, taken into the Forest; but Charford with its ninety-one acres of meadow-land, seems not to have been afforested, which, taken with other instances, shows that the best land was, as a rule, spared.

5. In the *Gentleman's Magazine* for 1828, vol. 98, part, ii., p. 17, is a sketch of the house, taken fifty years ago, which, with the

exception of some parts now pulled down, much resembles its present condition.

6. Monmouth, like a second Warbeck, was in all probability on his way through the Forest to Lymington, where Dore, the mayor, had raised for him a troop of men, and would assist him to embark. At Axminster, in Dorsetshire, there is a local MS. record, "*Ecclesiastica, or the Book of Remembrance,*" made by some member of the Axminster Independent Chapel, of the sufferings of Monmouth's followers, which appears to have been unknown to Macaulay.

7. There was formerly a cell here, subordinate to the Abbey of Saint Saviour le Vicomte in Normandy, to which it was given by William de Solariis, A.D. 1163, but dissolved by Henry VI., and its revenues annexed to Eton. Tanner's *Notitia Monastica*, Hants., No. xii. See, also, Dugdale's *Monasticon Anglicanum*, Ed. 1830, vol. vi., part, ii., p. 1046.

8. Same edition as before, p. iv. a. The entry is remarkably interesting. Out of its ten hydes, four were taken into the Forest. In the six which were left, there dwelt fifty-six villeins, twenty-one borderers, six serfs, and one freeman. There were here 105 acres of meadow, a mill which paid 22*s*., and a church with half a hyde of land. On the four hydes which were taken into the Forest, fourteen villeins, and six borderers, who had seven ploughlands, used to dwell. How very much the woodland preponderated over the arable we may tell by the additional entry, that the woods maintained 189 hogs, whilst a mill in that part was only assessed at 30*d*., which facts may help us to form some opinion of the kind of soil that was in general afforested. The meadows, as usual, were not touched.

CHAPTER XII.

THE VALLEY OF THE AVON CONTINUED.—TYRREL'S FORD, SOPLEY, AND WINKTON.

Tyrrel's Ford.

After we leave Ringwood the road for a mile or two is less attractive in its scenery. Still, here, as in every part of England, there is something to be seen and learnt. The Avon flows close by, famous for a peculiar eel, locally called the "sniggle" (*anguilla mediorostris*), which differs from its common congener (*acutirostris*) in its slender form and elongated under-jaw, and its habits of roving and feeding by day.[1] The river has, also, like some of the Norwegian streams, the peculiarity of forming ground ice. For the botanist, along the hedge banks, the blue and slate-coloured soapwort is growing throughout the summer and autumn, with purple cat-mint and wild clary. In the waste places the thorn-apple shows its white blossoms; whilst red stacks of fern and black turf ricks stand by every cottage door to remind us how close we are to the Forest.

After we pass Bisterne,[2] the road becomes more interesting. To our right rises the range of St. Catherine's Hills, that is, the fortified height, where remain the four mounds of the watchtowers and the traces of the camp. Presently we come to Avon-Tyrrel and the blacksmith's forge, built on the spot where Tiril's horse is said to have been shod, and which pays a yearly fine of three pounds and ten shillings to Government.

The actual Ford itself is some little way from the road. Round it stretch meadows, with strong coarse grass and sedgy weeds, branches of the Avon winding here and there, fringed by willows, the main stream flowing out broad and strong, with islands of osiers and rushes, where still breed wild duck and teal, the whole backed by the gloom of St. Catherine's Hills crested by their darker pines. The old road, used now only by the turf-cutters, crossing the former mill-brook, follows the bed of one of the many streams, till, reaching the river at its widest part, bends across, gaining a lane on the opposite side, which leads away past Ramsdown into Dorsetshire, and along which tradition says the knight rode to Poole.

The next village we reach is Sopley, that is the *soc leag*, land with the liberty of holding a court of *socmen*; just as the neighbouring village is called Boghamton (*bocland*)—the village of the charter-land, or, as we should now say, freehold. Its interesting little church, Early-English and Perpendicular, is dedicated to St. Michael, and built, in memory of the saint's burial-place, on a mound. The Avon flows below, and the old manor-house, now a mere cottage, stands in an adjoining meadow. On the deep north porch rests the archangel, on a corbel head. The fine old oak roof of the nave was covered up some sixty or seventy years ago by a plastered ceiling; but the corbel figures, playing the double pipe and viol, are still standing. In the north aisle are the heads of Edward III. and his queen. Two brackets for images project from the window in the north transept, whose jambs, now whitewashed over, were once painted with frescoes of the mystical vine, in green and red. Here, in the north wall, too, is an aumbrie, whilst the broken stone stairs to the rood-loft still remain. In the north transept a hagioscope looks into the chancel, where, on the floor, lie two Early Decorated figures, formerly placed in tombs under the rood-loft, and said originally to have been brought from the ruined church of Ripley. In the east window burns the fiery beacon of the Comptons.

Here, too, the whole of the church has been most impartially, and, I may add, successfully defaced. Everywhere has a snowstorm of whitewash fallen. I know not why we in these days should think that God delights in ugliness. Our forefathers at least thought not so. It would be well if for a moment we would consider how He adorns his own house, leads the green arabesque of ivy over its walls, and brightens the roof with the silver rays of mosses, and crowns each buttress with the aureole of the lichen.

Leaving Sopley, we come to "Winkton, the Weringetone of *Domesday*, where stood two mills, which were rated, as we have seen was often the case, by their yield of eels.

The views here are full of quiet beauty; the river winding along between its green walls of rushes, set with white and purple comfrey and yellow loosestrife, flowing into the darkness of the trees, and then again coming out by meadows, across which rises the Priory Church of Christchurch, standing out clear and sharp against the dark mass of Hengistbury Head.

The Avon at Winkton.

Footnotes

1. See Yarrell's *History of British Fishes*, vol. ii. pp. 399–401.
2. The ordnance map here falls into an error, placing Sandford a mile too far to the south; whilst it omits the neighbouring village of Beckley, the Beceslei of *Domesday*, and "The Great Horse," a clump of firs, so called from its shape, a well-known landmark in the Forest, and to the ships at sea, as also "Darrat," or "Derrit" Lane. For the barrows in the latter place, see chapter xvii.

CHAPTER XIII.

CHRISTCHURCH.

The Priory Church from the Castle Keep.

I HAVE determined to give a chapter to Christchurch, not because it contains more than many another town, but because it is a fair representative of the generality of small English boroughs. There is not a town in England, dating from even the Middle Ages, which is not full of interest peculiarly its own, and which does not possess memorials of the past which no other place can show. It has been proposed, by a no mean authority, to teach history by paintings and cartoons. But history is already painted for us on our city walls, and written for us upon our gates and crumbling castles. Our towns are in themselves the best texts upon history. For what we have seen with our eyes, and touched with our hands, leaves a more vivid and more lasting impression than the closest study of libraries of histories.

Further, the picture of a mediaeval town, as given in its own archives, with its own legislation, its peculiar manufacture, or import, forms, to some extent, the true social picture of the times. Its history reflects—and not faintly—the history of the day. Christchurch was never a town of sufficient importance to show all this in its municipal records. Yet, too, we shall see that they in another way are, like the town itself, full of interest. From a modern point of view there is nothing to be seen beyond three or four straggling streets and its manufactory of fusee watch-chains—the only one in England. All its interest and associations lie with the past. The country round it, too, is equally bound up with that same past. To the north rises St. Catherine's Hill, which we saw from the valley of the Avon, with its oval and square camps, and rampart and double vallum, crested with the mounds of its Roman watch-towers. The river Stour winds along between rows of barrows. Hengistbury Head is still fortified by its vast earthworks, and entrenched by deep ditches from the Avon to the sea.[1] Here the Britons saw the first swarm of fugitive Belgæ land and spread themselves along the rich valleys of Dorsetshire.[2] Here, centuries afterwards, the West-Saxons watched the raven-standard of the Danes scouring down the Channel, and knew their course along the coast, at night, by the blaze of burning villages, and, in the day, by the black trail of smoke.[3]

But to return to the town. Its Old-English names, Tweonea and Twinham-burn, were given to it from its situation between the rivers Avon and Stour. They were afterwards corrupted into the Norman Thuinam; which was lost in the name of its Priory, which overshadowed the town with its magnificence.

Here, in 901, came Æthelwald the Ætheling, son of Æthered, in his rebellion against his cousin Edward the Elder, and seized the place. From Christchurch he fell back upon Wimborne, which he fortified, exclaiming he would do one of two things, "Either there live, or there lie." That same night he fled to Northumberland.[4]

From *Domesday* we find that in Edward the Confessor's time the manor belonged to the Crown, and that thirty-one tenements paid a yearly tax of 16*d.* and a mill 5*s*, whilst another, belonging to the Church, was worth but 30*d.* Its woods, only, were enclosed in the Forest.

The manor remained in the hands of the Crown till Henry I. bestowed it on his friend and kinsman Richard de Redvers, Earl of Devon, the ruins of whose castle still overlook the Avon. Here his son Baldwin de Redvers in vain fortified himself against Stephen. Here, too, lived his grandson William de Vernon, who helped to bear the canopy at Richard's second coronation at Winchester. Afterwards, the manor passed into the hands of Isabella de Fortibus, who, on her death-bed, sold it, with all her possessions, to Edward I., who well knew the value of such a stronghold. Though Edward II. bestowed the estate on Sir William Montacute, yet the castle still remained in the hands of the Crown.

It was standing, though no longer a fortification, in the Commonwealth period. Nothing, however, now remains but the mere shell of the keep, whose walls are in places four yards thick. [5]

Below it stands what was, perhaps, the house of Baldwin de Redvers, also in ruins, and roofless, but still a capital specimen of what is so rarely seen, the true domestic architecture of the twelfth century. Like all the other remaining houses of this period, it is a simple oblong, seventy-one feet by twenty-four broad, and only two stories high, placed for defence on a branch of the Avon, which serves as a moat. On the south-east it is flanked by a small attached tower, now in ruins, under which the stream flows. The ground floor was divided in half by a wall, whilst the outer walls, thicker on the east and south sides, where more exposed to attacks than on the north and west, are pierced with deeply-splayed loopholes looking out on the stream. On this side was the hall, where lord, and guest, and serf, alike ate and drank, and slept on the floor. The other western half was divided into chambers and cellars, the kitchen probably standing in the courtyard.

The Norman House.

Above, approached by two stone staircases from within, and not, as in most cases, from without, was the principal dwelling-room, the solar, lighted on each side by three double lights, carved on their outer arches with zig-zag and billet mouldings, and on the south by a circular, and on the north by a fine double window, once richly ornamented, but now nearly destroyed. The fire-place, the only one in the house, is set nearly in the centre of the east wall; and above it still stands, in the place of the old smoke-vent, the beautiful round chimney, one of the earliest in England, like the fire-place, hid in ivy.

There seems, however, as in the case of the still older Norman house at Southampton, to have been no wall-passage connecting the building, as we might have expected, with the castle; but like it, its entrances, of which there were three, one opening out upon the stream, were on the ground floor.[6]

Coming down to later times, the great Lord Clarendon here possessed large property, and one of his favourite schemes was to make the Avon navigable to Salisbury. For this purpose it was surveyed by Yarranton, the hydrographer, who not only reported favourably of the idea, but proposed to make the harbour an anchorage for men-of-war, bringing forward the great natural advantages of Hengistbury Head, as also the facilities of procuring iron in the district, and wood from the New Forest.[7] All, however, fell to the ground with Clarendon's exile, and the harbour is now silted up with sand and choked with weeds.

Nothing else is there to be mentioned, except the visit by Edward VI. to the town, from whence he wrote a letter to his friend Barnaby Fitz-Patrick, far superior to most royal letters. The lazar-house, which stood in the Bargates, has long since been destroyed. The old market-place has been lately taken down; but in the main street, not far from the castle keep, remains, lately restored, one of those timbered houses so common in the midland counties and the Weald of Kent, with their dormer windows and richly-carved bressumers and barge-boards, but rarer in the West of England.[8] The glory, however, of the town, the Priory

Church, still stands. Before describing it let us give some account of its history. Its earliest buildings were founded by some of the secular canons of the order of St. Augustine, probably on a spot used for worship by the Romans.[9] Mention of it is made in *Domesday* as existing in Edward the Confessor's reign, and as possessing five hydes and one yardland in Thuinam, as also its tithes, and the third of those of Holdenhurst.[10] The present building, however, dates only from the time of Flambard, who rebuilt the church, pulling down the earlier building with its nine cells.[11] and introducing the regular instead of the secular canons. Not till Henry I.'s reign was the change completed, Baldwin de Redvers bringing in the former, and placing them under the first prior, Reginald.

With this change new privileges and grants were made. Riches flowed in on every side. Not only were the Redvers benefactors, but the Courtenays, and Wests, and Salisburys, into whose hands the manor of Christchurch came.[12]

Like most other ecclesiastical buildings, we hear but little of it till its dissolution. From its state we may be able to judge of the general condition of the monasteries, and how imperative was the change.

Leland[13] tells us that the Priory possessed but one volume—a small work on the Old-English laws. Their own accounts show us that the rules of St. Augustine had long been forgotten. Drunkenness had taken the place of fasting; and instead of giving they now owed.[14] Tradition, too, adds that the brethren were known in the town as the "Priory Lubbers." To this had the Austin Canons sank. So it was throughout England. Abbot and poorest brother were alike steeped in sensuality, and benighted in ignorance.

Of the last prior, John Draper, we catch some faint glimpse in a letter from Robert Southwell and four other commissioners to Cromwell, dated from Christchurch, the 2nd of December. He appears to have been a man who trimmed his course with the breath of authority, utterly selfish, utterly despicable. Not one word does he appear to have raised on behalf of his priory. Not one sigh did he utter for the old, nor one aspiration after the new religion. Thus the commissioners write:—"Our humble dewties observyd unto y^r gudde Lordeschippe. It may lyke the same to be advertised that we have taken the surrender of the late priorye of Christ Churche twynhm, wher we founde the prior a very honest, conformable pson, And the howse well furnysshede w^t Jewellys and plate, whereof som be mete for the $King^s$ majestie is use as A litill chalys of golde, a gudly lardge crosse doble gylt, w^t the foote garnyshyd w^t stone and perle, two gudly basuns doble gylt having the Kings armys well inamyld, a gudly great pyxe for the sacramet doble gylt, And ther be also other things of sylv, right honest and of gudde valewe as well for the churche use as for the table resyvyd, and kept to the Kings use."[15] Before the Dissolution came, whilst matters still trembled in the balance—whilst still there was hope that Protection would, for a little time longer, be given to hypocrisy, and Authority to sloth, he pleaded with Henry.[16] Now, when all hope was lost, when the end had arrived, the commissioners compliment him as the "very honest, conformable person." Had he previously been in earnest they must have written very differently. By his conformity he purchased his peace. And so, after giving up his priory, he was allowed to depart with a pension, to finish his life as he pleased, at the Prior's Lodgings at Sumerford Grange. There he died; and was buried in front of what had been his own choir; and his chantry still remains in the south choir aisle. Of the conventual buildings, which stood on the south side of the church, nothing remains except the fragments of the outer wall and the entrance lodge, built by Draper, with his initials still carved on the window label. A modern house stands on the site of the Refectory; and in digging its foundations, some tombs of the fourth century were found.[17] Other traces remain only in the names of the places, as Paradise Walk, by the side of the mill stream, and the Convent meadows, where, in an adjoining field, are the sites of the fishponds of the brethren.

The church stands at the south-west of the town, on a rising ground between the two rivers, its tower alike a seamark to the ships and a landmark to the Valley. But the first thing which strikes the visitor is not so much the tower, as the deep, massive north porch, standing right out from the main building, reaching to its roof, with its high-recessed arch, and its rich doorways dimly seen, set between clusters of black Purbeck marble pillars, and ornamented above with a quatrefoiled niche.

Standing here, and looking along the north aisle, the eye rests on the Norman work of the transept, the low round arches interlacing one another, their spandrels rich with billet and fishscale mouldings; whilst beyond rises the Norman turret, banded with its three string-courses, and enriched with its arcades, the space between them netted over with coils of twisted cables.

This is true Norman work, such as you can see scarcely anywhere else in England. And imagine what the church once was—a massive lantern-tower springing up from the midst, the crown of all this beauty.

Beyond all this lovely Romanesque work, rises the north choir aisle, with its quatrefoiled parapet, whilst above gleam the traceried windows of the choir, with their flying buttresses; and beyond them again stands the Lady Chapel, surmounted by St. Michael's loft, ugly and vile.

Entering, and standing at the extreme south-west end, we shall see the massive Norman piers rise in long lines, lightened by their columns, and relieved by their capitals, the spaces above each arch moulded with the tooth ornament. Above springs the triforium with its double arches, some of their pillars wreathed with foliage, the central shafts chequered in places with network, and woven over with tracery. Above that again runs the clerestory, now spoilt, whilst an open oak roof, hid by a ceiling, but once rich with bosses and carved work, encloses all.

To go into details. The porch and north aisle are Early-English, whilst a Norman arcade runs the whole length of the south aisle. The tower, and choir, and Lady Chapel, are Perpendicular, and the nave, as far as the clerestory windows, Norman.

Passing through the rich but mutilated rood-screen, which, however, sadly blocks up the way, we reach the choir, with its four traceried windows on either side, and clustered columns, from which springs its groined roof with bosses of foliage and pendants bright with gold, whilst the capitals of the shafts and the quatrefoils of the archivolts are rich with colour. The stalls are carved with grotesque heads and figures, like those in the Collegiate Church of the Holy Trinity, at Stratford-upon-Avon. Before us now stands the lovely reredos, illustrating the words of Isaiah,—"There shall come forth a rod out of the stem of Jesse, and a Branch shall grow out of his roots." Jesse sleeps at the bottom, his hand supporting his head, whilst David, with his fingers on his harp-strings, and Solomon, sit on each side, the vine spreading upwards, bearing its leaf and full fruit in Mary, to whose Son the Wise Men are offering their presents. Such is the screen, and had the execution been equal to the design, it would have been the finest in England. The carving seems, however, never to have been finished, and certainly in parts only to have been roughly cut by some inferior hand, and never to have received the last touches of the master-artist. Even now, in its present condition, it stands before those of Winchester and St. Alban's, inferior only to that of St. Mary's Overie.[18]

Passing on we come to the Lady Chapel, with its traceried roof. Under the east window are the remnants of another rich screen. The high altar, too, with its slab of Purbeck marble cut with five crosses, remains, whilst two recessed altar tombs to Sir Thomas West and his mother stand in the north and south walls.

But what we should especially see, both for its beauty and its interest, is the Chantry Chapel, built for her last resting-place by Margaret, Countess of Salisbury, mother of Cardinal Pole. It stands in the north choir aisle, its roof rich with arabesque tracery and carved bosses, telling a curious story in our English history. Attainted of treason the Countess was confined two years in the Tower before she suffered. When the day of execution came, she walked out on the fatal Tower Green; and still firm—still to the last resolute—refused to lay her head on the block. "So should traitors do," she cried, "but I am none;" and the headsman was obliged to butcher her as best he could.[19]

In the same letter before quoted from the Commissioners for the Suppression of Monasteries, dated from Christchurch, occurs this passage:—"In thys churche we founde a chaple and monumet curiosly made of cane [Caen] stone pparyd by the late mother of Raynolde pole for herre buriall, wiche we have causyd to be defacyd, and all the armys and badgis clerly to be delete."[20] To this day the vengeance of Henry's commissioners is visible, her arms being broken, and the bosses defaced, though her motto, "*Spes mea in Deo est,*" can still be read.

At the end of this aisle, under the east window, lie the alabaster effigies of Sir John Chydioke and his wife. The knight, who fell in the wars of York and Lancaster, wears his coat of mail, his head resting on his helmet, and his hands clasped together in prayer. At the western end, adjoining the north transept, stand two oratories with groined roofs, enriched with foliated bosses, whilst the capitals, from which the arches spring, are carved with heads.[21]

In the south choir aisle stand more monuments, amongst them the mortuary chapel of Robert Harys, with his rebus sculptured on a shield; and the chapel of Draper, the last prior, noticeable for its rich canopied niche over the doorway.[22]

And now that the reader has seen each part, let him go back to the west end, and sweep out of sight the whole thicket of pews, and break down the rood-screen blocking up the view, and looking through and beyond it, past the long line of Norman bays, with their sculptured tables, and past the chancel, imagine the stone reredos, as it once was, shining with gold and colour, all its niches filled with statues, and the windows above blazing with crimson and purple, through which the sunlight poured, staining the carved stalls and misereres,—and then he will have some faint idea of the former glory of the church.[23]

Most interesting is it, too, from another point of view. Since the Austin canons were more especially concerned with man's struggle in daily life, their churches assumed a parochial character. Hence we here have the spacious nave, so different to that of the old Nunnery Church of Romsey, the west tower and doorway—absent at Romsey—and the lovely north porch looking out to the town.

The whole building, I am sorry to add, is sadly out of repair. Restoration has been going on for some time past; but here, as in all similar cases, money is sadly needed. Surely men might give something, if from no higher motive than of keeping up a memorial of the piety of a past age. We inveigh against Cromwell and the Puritans—against the sacrilege of horses stabled in the choir, and the stalls turned into mangers;—against the sword which struck down the sculptured images, and the fire which consumed the carved woodwork. But the harm which the Puritans wrought is little compared to ours, in allowing the loveliness of our churches to rot by our negligence, and their sacredness to perish by our apathy.

The North Porch and Doorway.

Footnotes

1 In *Archæologia*, vol. v. pp. 337-40, is a description, illustrated with a plan of these entrenchments, together with the adjoining barrows, most of which have been opened, but the accounts are very scanty and unsatisfactory.
2. See Dr. Guest on the "Belgic Ditches," vol. viii. of the *Archæological Journal*, p. 145.
3. Gibson, in his edition of *The Chronicle*—in the "nominum locorum explicatio," p. 50, seems to think that Yttingaford, where peace was made between the Danes and Edward, was somewhere in the New Forest, deriving the word from Ytene, the old name of the district. Mr. Thorpe, however, in his translation of *The*

Chronicle, vol. ii. p. 77, suggests that it may be Hitchen.
4. *The Chronicle*, Ed. Thorpe, vol. i. p. 178. *Florence of Worcester*, Ed. Thorpe, vol. i. pp. 117, 118.
5. Grose, in his *Antiquities* (vol. ii., under Christchurch Castle), gives the following curious extract from a survey, dated Oct. 1656, concerning the duties of Sir Henry Wallop, the governor:—"Mem.: the constable of the castle or his deputy, upon the apprehension of any felon within the liberty of West Stowesing, to receive the said felon, and convey him to the justice, and to the said jail, at his own proper costs and charges; otherwise the tything-man to bring the said felon, and chain him to the castle-gate, and there to leave him. Cattle impounded in the castle, having hay and water, for twenty hours, to pay fourpence per foot." The fee of the Constable in the reign of Elizabeth was 8*l* 0*s*. 9*d*. Peck's *Desiderata Curiosa*, vol. i., book ii., part. 5, p. 71. In the Chamberlain's Books of Christchurch we are constantly meeting with some such entry as, "1564, ffor the castel rent for ij yeres—xiij*s*. v*d*." "1593, ffor the chiefe rent to the castel—vi*s*. xi*d*.
6. Descriptions of it will be found in Hudson Turner's *Domestic Architecture of England*, vol. i. pp. 38, 39. Parker's *Glossary of Architecture*, vol. i. p. 167. Grose's *Antiquities*, vol. ii. Hampshire; in whose time it appears to have been cased with dressed stones. In the Chamberlain's Books of the Borough, under the date of the sixth year of Edward VI., 1553, we meet with repairs "for the house next the castle," which entry probably refers to some buildings belonging to the house, which, according to Grose, stretched away in a north-westerly direction to the castle.
7. *England's Improvements by Sea and Land*. By Andrew Yarranton. Ed. 1677, pp. 67. 70.
8. As we have said, the muniment chest of the Christchurch Corporation, like that of all similar towns, is full of interest. It contains absolutions from Archbishops to all those who assist in the good work of making bridges;—letters from absolute patrons

directing their clients which way to vote;—bonds from others that they will not require any payment from the burgesses, or put the borough to any expense;—old privileges of catching eels and lampreys with "lyer," and "hurdells de virgis," by all of which the past is brought before us. So, too, the Chamberlain's Books are most interesting. From them we can learn, year by year, the prices of wheat and cattle, the fluctuation of wages, the average condition of the day, and both the minutest outward events as also the innermost life of the town. The true social history of England is written for us in our Chamberlain's Books. They have unfortunately never been made use of as they deserve. Thus let me give a few general quotations from those of Christchurch. In 1578 lime was 6*d*. a bushel, from which price it fell within two years to 2*d*. Stone for building we find about 1*s*. a ton. Wages then averaged, for a skilled mechanic, from 7*d*. to 1*s*. a day, and for a labourer, 4*d*.; whilst night-watchmen, in 1597, were only paid 2*d*. Timber, contrary to what we should have expected, was comparatively dear. Thus in 1588 we find 9*d*. paid for two posts, and 20*d*. for a plank and two posts, whilst a few years afterwards a shilling is paid for making a new gate. Of course in all these calculations we must bear in mind that money was then three times its present value. Turning to other matters, we learn that in 1595, "a pottle of claret wine and sugar" cost 2*s*., whilst a quart of sack is only 12*d*. In 1582, a quart of "whyte wine" is 5*d*., and twenty years before this a barrel and a half of beer cost 4*d*. Again, in 1562, the fourth year of Elizabeth, large salmon, whose weights are not specified, appear to have averaged 7*d*. a piece. A load of straw for thatching came to 2*s*. 6*d*., and in some cases 3*s*., which in 1550 had been as low as 8*d*., and never above 20*d*. Drawing it, or passing it through a machine, cost 4*d*.; whilst a thatcher received 1*s*. 4*d*. for his labour of putting it on the roof.

At the same time a load of clay, either for making mortar or for the actual material of the walls, the "cob," or "pug" of the provincial dialect, was *5d.*, a price at which it had stood with some slight variations for many years.

To conclude, the smallest things are noted. Thus a thousand "peats," perhaps brought from the Forest, cost, in 1562, 15*d.*, whilst a load of "fursen," still the local plural of furse, perhaps also from the same place, was 8*d.* Nothing in these accounts escapes notice. In 1586 a "coking stole," the well-known *cathedra stercoris*, the Old-English "*scealfing-stol*," is charged 10*d.*; whilst a collar, or, as it is elsewhere in the same book called, "an iron choker for vagabonds," cost 14*d.*

9. *In Archæologia*, vol. iv. pp. 117, 118, is a letter from Brander, the geologist and antiquary, describing a quantity of spurs and bones of herons, bitterns and cocks, found on a part of the monastic buildings, showing that the site had been previously occupied.

10. Holdenhurst had ten hydes and a half taken into the Forest (*Domesday*, as before, iv. a). It then possessed a small church, and, as we find one mentioned in the charter of Richard de Redvers in Henry I.'s reign, we may fairly conclude that this, too, was not destroyed by the Conqueror.

11. *Cartularium Monasterii de Christchurch Twinham*. Brit. Mus., Cott. MSS., Tib. D. vi., pars ii., f. 194 a. This chartulary was much injured in the fire of 1731, but has been restored by Sir F. Madden. Quoted in Dugdale's *Monasticon Anglicanum*, vol. vi. p. 303, Ed. 1830.

12. For further information, especially on the fortunes of the De Redvers family, and minor details, which I hardly think would interest the general reader, see Brayley's and Ferrey's work on the Priory of Christchurch, London, 1834, pp. 6. 11. 22: and Warner's *South-west Parts of Hampshire*, vol. ii. pp. 55-65, which, notwithstanding some errors, is a most painstaking history.

13. *Collectanea de Rebus Britannicis*, Ed. Hearne, vol. iv. p. 149.

14. The possessions of the house were large, and brought in above 600*l.* a year. Yet we find that the brethren were in debt in every

direction. At Poole, Salisbury, and Christchurch, they owed 41*l*. 19*s*. 6*d*. for mere necessaries. There was due 24*l*. 2*s*. 8*d*. to the Recorder of Southampton for wine; and a bill of 8*l*. 13*s*. 2*d*. to a merchant of Poole, for "wine, fish, and bere." Certificate of Monasteries, No. 494, p. 48. Record Office. Quoted by Brayley and Ferrey, Appendix No. vi., pp. 9, 10.

15. Brit. Mus., Bibl. Cott., Cleopatra, E. iv., f. 324 b.

16. "Petition of John Draper." Amongst the Miscellaneous MSS. of the Treasury of the Exchequer, Record Office.

17. *Archæologia*, vol. v. pp. 224-29.

18. I know nothing equal to this last screen in the delicacy of its carving, seen in bracket, and canopy, and the flights of angels: in the deep feeling especially manifest in the central bracket, with the Saviour's head crowned with thorns, but surrounded with fruit and flowers, typical of His sufferings and the world's benefits; and in the grave humour, not out of place, as allegorical of the world's pursuits, which peeps forth in the figures over the two doorways.

19. Lord Herbert's *Life and Reyne of King Henry VIII.*, Ed. 1649, p. 468. See, however, Froude's *History of England*, vol. iv. p. 119, foot-note.

20. The year, as was generally the case, is not given to this letter, but simply December 2nd. From internal evidence, however, it was certainly written in 1539; for we know that the Priory was surrendered Nov. 28th of that year. Why, then, two years before her death, the commissioners should speak of the "late mother of Raynolde pole" I know not.

21. Below the north transept, part, perhaps, of Edward the Confessor's church, is a vault, which, when opened, was stacked with bones, like the carnary crypts at Grantham, in Lincolnshire, and of the beautiful church at Rothwell, in Northamptonshire—the "skull houses," to which we so often find reference in the old churchwardens' books.

22. In the south choir aisle the broken sculptures represent the Epiphany, Assumption, and Coronation of the Virgin. Little can be said in praise of any of the modern monuments. The best are Flaxman's "Viscountess Fitzharris and her three Children," and Weekes's "Death of Shelley." Some of the others should never have been permitted to be erected, especially those which disfigure the Salisbury chapel. The new stained window at the west end adds very much to the beauty of the church.
23. For further details the student of architecture should consult Mr. Brayley and Mr. Ferrey's work, before referred to, of which a new edition is much needed, as also Mr. Ferrey's paper in the *Gentleman's Magazine* for Dec., 1861, p. 607, on the naves of Christchurch and Durham Cathedral, both built by Flambard, and a paper on the rood screen in the *Archæological Journal*, vol. v. p. 142; and also a paper read at Winchester, September, 1845, before the Archæological Institute, on Christchurch Priory Church, by Mr. Beresford Hope, and published in the Proceedings of the Society, 1846. An excellent little handbook, by the Secretary of the Christchurch Archæological Association, may be obtained in the town.

CHAPTER XIV.

THE OLD SOUTH-WESTERN SEA-COAST — SOMERFORD, CHEWTON GLEN, MILFORD, HURST CASTLE, LYMINGTON.

Chewton Glen.

LITTLE has been seen of the sea, except from Calshot Castle to Leap. Though, too, the sea-coast here, as there, is no longer in the Forest, yet if we miss this walk we shall lose some of the most beautiful scenery in the district.

As we leave Christchurch by the Lymington Road, Mudeford lies on the right, and Burton, with its Staple Cross, on the left. Few things are more touching than these old grey relics of the past, standing solitary in our cross-roads, the dial united with the Cross, to show both how short was man's life, and where lay his only salvation. But we now profane them, and turn them, as here, into direction posts, or break them up, as at Burgate, to mend the road.

Both villages will some day be more sought after than at present, for at Burton lived Southey, with his friend Charles Lloyd, and sang the praises of the valley in better verse than usual. At Mudeford, Stewart Rose, the author of *The Red King*, built Gundimore, where, in 1807, Scott stayed, writing *Marmion*, and riding over the Forest exploring the barrows. In the same village Coleridge lodged during the winter of 1816.[1]

A little way along the main road lies Somerford, once one of the Granges of Christchurch Priory. Its barns and stables are partly built from the prior's lodgings, whose site may here and there be faintly traced; and the chapel, which in Grose's time was still standing, with the initials of the last prior, John Draper, cut on the window labels.[2]

The best plan, however, is not to go along the road, but the shore as far as Chewton Glen, and there climb up the cliff. The sands are white and hard, strewed with fragments of iron-stone, and large *septaria*, from which cement is made, and for which, farther on, a fleet of sloops are dredging a little way from the shore. In the far distance gleam the white and black and orange-coloured bands of sand and clay scoring the Barton cliffs.[3]

The glen, or "bunny," as it is locally called, runs right down into the sea; the high tide rushing up it, and driving back its Forest stream. Down to the very edge it is fringed with low oak copses, covered in the spring, as far as high-tide mark,

THE OLD SOUTH-WESTERN SEA-COAST OF THE NEW FOREST.

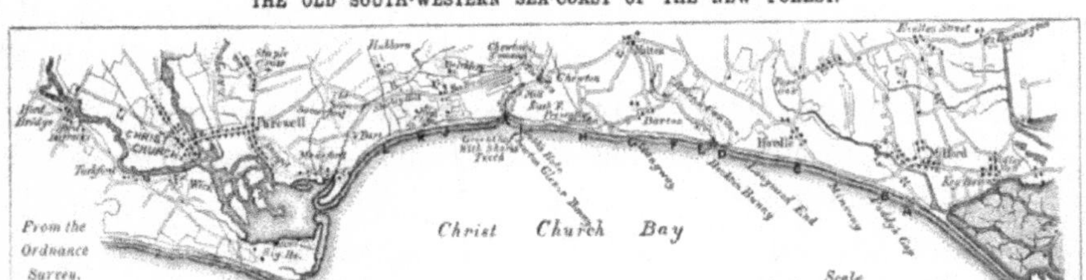

A.—Middle Marine Bed.
B.—Fresh-water Clays and Marls.
C.—Crocodile Bed, &c., exposed.
D.—Olive Bed, well exposed.
E.—Lignite Bed at the top of Cliff.
F.—Chama Bed.
G.—Barton Clay. Fossils abundant
H.—Green Clay, with sharks' teeth, and bones of fish.
I.—High Cliff Sands rise.
J.—Pebble Bed.
K.—Grey Sands, interstratified with fossil-wood and iron-stone
L.—Bracklesham Sands.

with blue bells, and strewed with yellow tufts of primroses. In the summer, too, the ground is as deep a green with ferns as the oak leaves above; whilst the stream flows between banks bordered with blue skullcap and purple helleborine.[4]

Then, as you climb up to the down, on the opposite side, stretches a view, hard to be matched in England either for extent or beauty. On one side rolls the English Channel, indenting the shore with its deep bay as far as the land-locked harbour of Christchurch, shut in by Hengistbury Head and the white Swanage rocks; and, on the other, it sweeps away by the long beach of Hurst and its round gray castle. Opposite, glitter the coloured sands and chalk cliffs of Alum Bay, and the white Needle Rocks running wedge-shaped into the sea. Farther eastward, rise the treeless downs, and the breach opens across the Island to Freshwater Gate, and the two batteries, built into the cliff, one by one appear: the long scene ended at last by the houses of Yarmouth—the Solent still winding onward, like some great river.

An uninterrupted path runs, for some three or four miles, along the top of the cliff—the scene constantly changing in its beauty. Below hangs a broken under-cliff, shelving down to the sea, strewed here and there with blocks of gravel, the grass and furze growing on them just as they fell. On the shore stretch long reaches of yellow sand, separated by narrow strips of pebbles, and patches of dark green Barton clay, embossed with shells, and studded with sharks' teeth.

Passing the Coastguard Station and the Gangway, we reach Becton Bunny—very different to Chewton, but equally lovely, with its bare wide gorge, and its beds of furze and heath fringing the edge of the cliff.[5] Very beautiful, too, are the summer sunsets seen from this point—the sun sinking far down the channel, lighting up the coloured sands of Alum Bay purple and gold, tinting the white chalk cliffs with rose and vermilion, the crimson of the sky floating on the waves as they break along the shore.

Still following the path along the top of the cliff, we pass the grave-yard, where stood the old cruciform church of Hordle—once in the middle of the village, but now only a hundred yards from the sea. Nothing of it remains except some blocks of Grey Wethers, used for its foundation, and too large to be removed. Very interesting are these stones, brought up from the shore, where, now and then, one or two may be seen at low tide, tumbled from the drift above—the same stones as those at Stonehenge, left on the top of the chalk. Gone, too, are its mill and its six salterns, mentioned in *Domesday*, and the village itself removed inland. The sailors, however, dredging for cement-stone or for fish, sometimes draw up great logs of wood, locally known as "mootes," which may perhaps tell of the salterns, or the time when the Forest stretched to the sea. The salterns of the Normans and the Old-English have suffered very different fates. In Normandy the sea no longer reaches to their sites,[6] whilst here it has long since rolled over them.

Beyond this again is Mineway, reminding us, by its name,

of the time when the iron-stone was collected on the shore and taken to the Sowley furnaces to be smelted.[7] Farther on, down in the valley made by the stream, which turns the village mill, mentioned in *Domesday*, lies Milford. The church spire rises up prettily amongst its trees, and the church itself is a good example of our village churches, built in three or four different styles. The tower is Early-English, surmounted by a string-course of Norman heads. In the north side stands a curious inserted doorway, with trefoil heading, whilst two Norman arches remain in the nave joined by Early-English, springing from black Purbeck marble shafts.

To the south stretches the long Hurst beach, formed, in much the same way as the more famous Chesil Bank, of the rolled pebbles brought up from the Barton Cliffs by the strong tides aided with the westerly gales, making a breakwater to the whole of the Solent. Now and then close to it appear the floating islands, known as the Shingles, sometimes rising for only a few hours above the sea, and at others remaining long enough to become green with bladderwort and samphire.

Across to the Isle of Wight, at the narrowest point, it is only a mile; and so fast does the Solent tide,[8] when once the ebb is felt, pour itself along the narrow gorge, that it fills up Christchurch Bay, higher than at the flood, thus making, in fact, a double high-water. At the extreme end stands Hurst Castle, built by Henry VIII., from the ruins of Beaulieu Abbey. Whatever opinion we may have of Henry's private character, there can be but one as to his foresight and energy in defending the country. Much for this may be forgiven. Hall wrote in no exaggerated strain when he said:—"The King's highness never ceases to study and take pains both for the advancement of the commonwealth of this his realm of England, and for the defence of the same. . . . . Wherefore, his Majesty in his own personne took very laborious and painful journeys towards the sea-coasts. Also, he sent dyvers of his nobles and counsellors to view and search all the portes and dangers in the coastes, ... and in all soche doubtful places his Highness caused dyvers and many bulwarks and fortifications to be made."[9] And of them, Hurst Castle, like Calshot, which we have seen, was one, and still stands, additionally fortified by guns, and guarded by the far better defences of lighthouses, and beacons, and telegraph stations.[10]

Here it was, on the 1st December, 1642, Charles I. was brought, after holding his mock court at Newport, by Colonel Cobbit, who had seized him in the name of the army. Here, too, he still showed all the foolish childishness which Laud had taught him, putting faith in the omen of his candle burning brightly or dimly,[11] which detracts so much from any interest we might otherwise feel for him in his days of care and sorrow. A closet is shown where he is said to have been confined, and where his Golden Rules are said to have hung; but from Herbert's memoirs, evidently neither the room where he lived or slept.[12] Herbert's account of Hurst is so graphic that I give it nearly in full:—"The wind and tide favouring, the King and his attendants crossed the narrow sea in three hours,[13] and landed at Hurst Castle, or Block House rather, erected by order of King Henry VIII., upon a spot of earth a good way into the sea, and joined to the firm land by a narrow neck of sand, which is covered over with small loose stones and pebbles; and upon both sides the sea beats, so as at spring tides and stormy weather the land passage is formidable and hazardous. The castle has very thick stone walls, and the platforms are regular, and both have several culverines and sakers mounted. . . . The captain of this wretched place was not unsuitable; for, at the King's going ashore, he stood ready to receive him with small observance. His look was stern. His hair and large beard were black and bushy. He held a partizan in his hand; and, Switz-like, had a great basket-hilt sword on his side. Hardly could one see a man of more grim aspect, and no less robust and rude was his behaviour."[14] The account is very life-like, though some allowance must be made for Herbert's prejudices against this gaunt Puritan captain, who, we learn, by-and-by became more civil. Colonel Cobbit, in whose charge the King was, seems to have treated him with uniform respect and kindness. Charles stayed here six-and-twenty days, walking along the beach, watching the ships passing up and down the Solent, and receiving the cavaliers of Hampshire, who came for the last time to pay their respects. Then, at last, he was suddenly taken away to show at Whitehall a better courage and wisdom in death than in life.

About three miles from Milford, on the mouth of the Boldre Water, lies the port of Lymington, the Mark of the Limingas, as the neighbouring hamlet of Pennington is that of the Penningas.[15] Its manor, like that of Christchurch, once belonged to Isabella de Fortibus, and was given, with some other possessions, by Edward I., to her rightful heir, the Earl of Devon, whose arms are still quartered with those of the Corporation. It is another of those towns, which, like Christchurch, though in a very different way, is associated with the past. It has no monastic buildings, no ruins of any kind, no church worth even a glance. Yet, too, it can tell of departed greatness.

From the coins which have been dug up in the town, and the camp at Buckland Rings,[16] it was evidently well known to the Romans. In *Domesday*, the famous Roger de Yvery held one hyde here; but its woods were thrown into the Forest, and for this reason the manor was only rated at one half. No mention is made of its salt-works, though we know, from a grant of Richard de Redvers, in 1147, confirming his father's bequest of the tithe of them to Quarr Abbey, that they were then probably in existence.[17] Larger than Portsmouth, in 1345, it contributed nearly double the number of ships and men to Edward III.'s fleet for the invasion of France. We must not, however, conclude that it has decreased.

[18] Larger now than ever, like so many other old towns, it has not increased in a relative proportion with younger rivals favoured by the accidents of position or commerce. Like, too, all other similar ports, it has its tales to tell of French invasions, and, like similar boroughs, of the Civil War; but they are merely traditional, and, therefore, vague and unsatisfactory. Loyal from first to last, it is said to have at its own cost supplied with provisions the ships of Prince Charles, when he lay in the Yarmouth Roads, hoping to rescue his father from Carisbrook. In still later times, carried away by Protestant sympathies, it espoused the cause of the imbecile Monmouth, the mayor raising some hundred men to join his standard.[19]

Most of the places round Lymington, Buckland Rings, Boldre Church, Sway Common, with its barrows, we have already seen. A little, though, to the eastward, at Baddesley, near Sowley Pond, formerly stood a Preceptory of the Knights Templar, and afterwards of those of St. John of Jerusalem. At the Dissolution it was granted to Sir Thomas Seymour, and again by Edward VI. to Sir Nicholas Throckmorton, but subsequently, under Mary, restored to the Hospitallers. Nothing of it is now left.[20]

Here, then, at Lymington, we have been the whole circumference of the Forest. I do not know that I have omitted anything of real interest. Mere idle gossip, vague stories, I have left to those who care to write, and those who like to read such things. The geology, and botany, and folk-lore of the district, to which it was impossible to do more than to make general references, will be found in the succeeding chapters. As was before said, in the wild commons and woods themselves I have myself taken the greatest interest, and wished to impress their beauty on the reader, feeling that a love for Nature is the mainspring of all that is noble in life, and all that is precious in Art. I do not know either that I have anywhere exaggerated. On the contrary, no words can paint, much more exaggerate, the loveliness of the woods. And of all walks in the district, this over the Hordle and Barton Cliffs is by no means the least beautiful, though no longer in the Forest.

Hurst Castle.

Footnotes

1. Scott used to admire the *Red King*; but I suspect, judging from quotations, his praise was rather the result of friendship than of unbiassed criticism. The following lines, from Rose's MS. poem of "Gundimore" (quoted in Lockhart's *Life of Scott*, p. 145, foot-note), are interesting from their subject, and at the conclusion, though the idea is borrowed, are really fine:—

> "Here Walter Scott has wooed the Northern Muse,
> Here he with me has joyed to walk or cruize;
> And hence has pricked through Ytene's holt, where we

Have called to mind how under greenwood tree,
Pierced by the partner of his 'woodland craft,'
King Rufus fell by Tiril's random shaft.
Hence have we ranged by Keltic camps and barrows,
Or climbed the expectant bark, to thread the Narrows
Of Hurst, bound westward to the gloomy bower
Where Charles was prisoned in yon island tower.

Here, witched from summer sea and softer reign,
Foscolo courted Muse of milder strain.
On these ribbed sands was Coleridge pleased to pace
Whilst ebbing seas have hummed a rolling base
To his rapt talk."

2. *Antiquities*, vol. ii., where there is a sketch of the Grange as it was in 1777.

3. For the geology of High Cliff, Barton, and Hordle Cliffs, see chapter xx. There are not many fossils in either the grey sand or the green clay before you reach the "bunny." Plenty, however, may be found in the top part of the bed immediately above, known as the "High Cliff Beds," and which rise from the shore about a quarter of a mile to the east of the stream.

4 Chewton is not mentioned in *Domesday*. Beckley (Beceslei), which is close by, where there was a mill which paid thirty pence, had a quarter of its land taken into the Forest; whilst Baishley (Bichelei) suffered in the same proportion. Fernhill lost two-thirds of its worst land, and Milton (Middeltune) its woods, which fed forty hogs, by which its assessment was reduced to one-half.

5. At this point the Marine Beds end, and the Brackish-Water series crop up; and then, lastly, the true Fresh-Water shells commence—the Paludinæ and Limnææ, with scales of fish, and plates of chelonians, and bones of palaeotheres, and teeth of

dichodons. See, further, chapter xx.

6. See Lappenberg's *England under the Anglo-Norman Kings.* Ed. Thorpe, p. 89.

7. Yarranton, in that strange but clever work, *England's Improvement by Land and Sea* (Ed. 1677, pp. 43-63), dwells at length on the quantity of iron-stone along the coast, and the advantage of the New Forest for making charcoal to smelt the metal. He proposed to build two forges and two furnaces for casting guns, near Ringwood, where the ore was to be brought up the Avon.

> 8. "That narrow sea, which we the Solent term,
> Where those rough ireful tides, as in her straights they meet,
> With boisterous shocks and roars each other rudely greet;
> Which fiercely when they charge, and sadly when they make retreat,
> Upon the bulwark forts of Hurst and Calshot beat,
> Then to Southampton run.
> — *Polyolbion*, book ii.

9. Hall's *Union of the Families of Leicester and York*, xxxi. year of King Henry VIII., ff. 234, 235; London, 1548.

10. From Peck (*Desiderata Curiosa*, vol. i., b. ii., part iv., p. 66) we find that in Elizabeth's reign the captain received 1*s.* 8*d.* a day; the officer under him, 1*s.*; and the master-gunner and porter, and eleven gunners and ten soldiers, 6*d.* each, which in Grose's time had been increased to 1*s.* (Grose's *Antiquities*, vol. ii., where a sketch is given of the castle). Hurst, on account of its strength, was to have been betrayed, in the Dudley conspiracy, to the French, by Uvedale, Captain of the Isle of Wight. (Uvedale's Confession, *Domestic MSS.*, vol. vii., quoted in Froude's *History of England*,

vol. vi. p. 438.) Ludlow mentions the great importance of Hurst being secured to the Commonwealth, as both commanding the Isle of Wight and stopping communication with the mainland (*Memoirs*, p. 323). Hammond, in a letter from Carisbrook Castle, June 25th, 1648, says it is "of very great importance to the island. It is a place of as great strength as any I know in England" (Peck's *Desiderata Curiosa*, vol. ii., b. ix., p. 383).

11. Sir Thomas Herbert's *Memoirs of the two last Years of the Reign of King Charles I.*, Ed. 1702, pp. 87, 88.

12. Warwick calls the King's rooms "dog lodgings" (*Memoirs*, p. 334); but it is evident from Herbert (*Memoirs*, p. 94) that both Charles and his attendants were well treated, which we know from Whitelock (*Memorials of English Affairs*, p. 359; London, 1732) was the wish of the army, as also from the letter of Colonel Hammond's deputies given in Rushworth (vol. ii., part iv., p., 1351). Of Colonel Hammond's own treatment of the King we learn from Charles himself, who, besides speaking of him as a man of honour and feeling, said "that he thought himself as safe in Hammond's hands as in the custody of his own son" (Whitelock, p. 321).

13. Evidently a misprint for three-quarters of an hour.

14. Herbert's *Memoirs*, pp. 85-86.

15. A Keltic derivation for both places has been proposed, but it is not on critical grounds satisfactory.

16. Gough possessed a brass coin inscribed Tetricus Sen. rev. Lætitia Augg., found here; and adds that in 1744 nearly 2 cwt. of coins of the Lower Empire were discovered in two urns. Camden's *Britannia*, Ed. Gough, vol. i. p. 132.

17. The grant is given in the Appendix to Warner's *South-West Parts of Hampshire*, vol. ii., p. i., No. 1.

18. Like those of Christchurch, the Corporation books of Lymington are full of interest, though they do not commence till after 1545, the previous records being generally supposed

to have been burnt by D'Annebault in one of his raids on the south coast. Du Bellay, however, who, in his *Mémoires*, has so circumstantially narrated the French movements, says nothing of Lymington having suffered, nor can I find it mentioned in any of the State papers of the time. Take, for instance, the following entries from the Chamberlain's books:—

"1643.	Quartering 20 soldiers one daie and night, going westward for the Parliam' service	xvi.*s*. ij.*d*.
1646.	For bringinge the toune cheste from Hurst Castell	ij.*s*.
1646.	Watche when the allarme was out of Wareham	iiij.*s*
1646.	For the sending a messenger to the Lord Hopton, when he lay att Winton with his army, with the toune's consent	xiiij.*s*.
1648.	For keeping a horse for the Lord General's man	iij.*s*. x.*d*.
1650.	Paid to Sir Thomas Fairfax his souldiers going for the isle of Wight with their general's passe	xij.*s*."

Such entries to an historian of the period would be invaluable, as showing not only the state of the country but of the town, when the town-chest had to be sent four miles for safety; and proving, too, that here (notice the fourth entry), as elsewhere, there were two nearly equally balanced factions—one for the

King, the other for the Commonwealth. I may add that a little book has been privately printed, of extracts from the Lymington Corporation books, from which the foregoing have been taken. It would be a very good plan if those who have the leisure would render some such similar service in other boroughs.

19. Warner's *Hampshire*, vol. i., sect, ii., p. 6; London, 1795. See, too, previously, ch. xi., p. 122, foot-note.

20. See Dugdale's *Monasticon Anglicanum*, vol. vi., part ii., p. 800. Tanner's *Notitia Monastica*. Ed. Nasmyth, 1787. Hampshire. No. iv.

CHAPTER XV.

THE GYPSY AND THE WEST-SAXON.

View in Mark Ash.

MANY people have a vague notion that the gipsies constitute the most important element of the population of the New Forest, whereas, of course, they are mere cyphers. An amusing enough French author, in a work upon England, has devoted a special chapter to the New Forest, and there paid more attention to the gipsies than any one else, and entirely forgets the West-Saxon, whose impress is indelibly marked, not only in the language, but in the names of every town, village, and field.

As, however, every one takes a romantic interest in these nomads, we must not entirely pass over them. Here and there still linger a few in whose veins run Indian blood, against whom Henry VIII. made bad laws, and Skelton worse rhymes. The principal tribes round Lyndhurst are the Stanleys, the Lees, and Burtons; and near Fordingbridge, the Snells. They live chiefly in the various droves and rides of the Forest, driven from place to place by the policeman, for to this complexion have things come. One of their favourite halting-places is amongst the low woods near Wootton, where a dozen or more brown tents are always fluttering in the wind, and as the night comes on the camp-fires redden the dark fir-stems.

The kingly title formerly held by the Stanleys is now in the possession of the Lees. They all still, to a certain extent, keep up their old dignity, and must by no means be confounded with the strolling outcasts and itinerant beggars who also dwell in the Forest. Their marriages, too, are still observed with strictness, and any man or woman who marries out of the caste, as recently in the case of one of the Lees, who wedded a blacksmith, is instantly disowned. The proverb, too, of honour among thieves is also still kept, and formal meetings are every now and then convened to expel any member who is guilty of cheating his kinsman.

Since the deer have been destroyed in the Forest, life is not to them what it was. They are now content to live upon a stray fowl, or hedgehog, or squirrel, baked whole in a coat of clay, and to gain a livelihood by weaving the heather into mats, and brooms, and beehives.

They are, however, mere wanderers, and have nothing to do with the soil. It is with the West-Saxons that we are most concerned. And in the New Forest he will be found just such another man as his forefather in the days of William the Red, putting the same faith in visions and omens which made the King exclaim, on the morning of his death, upon the news of the monk of Gloucester's dream, "Do you take me for an Englishman?" believing firmly in groaning ash-trees, and oaks which bud on Christmas-eve, and witches who can turn themselves into hares, deeming that the marl which he digs is still red with the blood of his ancient foes the Danes.[1]

Here, as we have seen in Hampshire, at Calshot, on the borders of the Forest, Cerdic landed. Here he defeated the Britons, and established the kingdom of the West-Saxons. Here the West-Saxon Alfred rallied his countrymen and crowned defeat with victory. Here, too, stood the capital of Wessex, Winchester, in whose cathedral lie the old West-Saxon kings. Here, then, if anywhere, we should expect to find West-Saxon characteristics and a West-Saxon population.

As is well known, after the battle of Hastings, the West-Saxons, with one or two exceptions, succumbed willingly enough to the Conqueror, who lived amongst them; whilst the Northmen across the Humber bid him defiance. Every one must to this day notice the extreme deference, almost amounting to a painful obsequiousness, of the lower classes in the southern, compared with their independent manner in the northern, parts of England. We find, too, mingled, however, with characteristics from other sources, the West-Saxon element not only in the appearance of the long-limbed Forest peasantry, with their narrow head and shoulders, and loose, shambling gait, but also in their slowness of perception. They betray, too, to this hour that worst Teutonic trait of fatalism, observable in all their epitaphs, and in their daily expression, "It was not to be," applied to anything which does not take place. Notwithstanding, too, their apparent servility, an amount of cunningness and craft peeps out, which in a different age compelled the Conqueror to make special laws against assassination.[2]

Much must be set against these drawbacks. Enslaved to an extent which no modern historian has dared to reveal, and can only be fully conceived by the dreadful story of *The Chronicle*,—treated as beasts rather than even slaves,—the West-Saxons showed, under the Normans, a spirit of obedience and an adaptability to changed circumstances which are above praise. Let us give the West-Saxon labourer credit for it both then and to this day, that though the most ill-paid and ill-fed in England, he bears his heavy yoke of poverty without a murmur.

Turning to another side of his character, we find him loving the same old sports as in the days of Alfred. He still follows the hounds on foot, and when there were deer in the Forest, naturally killed them. Wrestling and cudgel-playing have been continued till the last few years close to the northern boundaries of the Forest. The old Hock-tide games were till a late period kept up in the northern parts, and "Hock-tide money" was not so very long ago paid as an acknowledgment for certain Forest privileges. Heartiness and roughness still go hand in hand with him as with his forefathers. But a heaviness of intellect is always visible, and, as with all his race, a sadness oppresses his mirth. His dress to this day, too, bespeaks his nationality. He still wears what is locally called the "smicket," and sometimes the "surplice," the Old-English smoc, called also the tunece. It is still, too, as formerly, tied round the waist with a leathern band. His legs are still cased, as we see the Old-English in their drawings, with gaiters, known as "vamplets," or "strogs," equivalent to the "cockers" of the Midland Counties, which do not reach quite so high as the former, and "mokins," which are merely made of coarse sacking.

And now let us see how far he has made his presence felt on the district and in the language. But we must beware of overstraining our theory. No portion of our history is, in its details, so difficult as the English Conquest. None, to any statement which may be made, requires so many qualifications. The first faint flow of the Teutonic immigration was felt long prior to Caesar's invasion—centuries before the main wave burst over the country. We must, too, carefully bear in mind that in Wessex, more than in any other part, the conquerors and conquered were blended together. [3] They mixed, however, everywhere far more than is commonly allowed. Our language bears testimony to the general fact. The many Keltic household words in daily use are the best evidence.

Here in the New Forest I may mention that the form "plock" is used instead of the common block (*bloc*), and that we have, as, perhaps, throughout the West of England, "hob," in the sense of potato-hob — a place where potatoes are covered over, instead of "hog" (*hwg*), noticed by Mr. Davies in his list of Keltic words in Lancashire. Further, we find the terms "more" (*maur*), for a root, "mulloch," for dirt, and "bowerstone," for a boundary-stone.[4] Here, too, as in other places, the Britons have left the traces of their rule on the broader natural features of the country—on the rivers, as the Exe (*y* [*g*] *wysg*, the current), and Avon (*Afon*, the river), and Avon Water, near Setthorns, and Boldre (*y Byldwr*, the full stream), and Stour ([*G*]*wys-dwr*, the deep water), and in the district itself, in the now almost forgotten name of Ytene. We find their influence, too, perhaps, in such local names of villages and fields as Penerley, Denny, Cocketts, Kamel or Cammel Green, and Flasket's Lane. As might be expected, the traces of the Danes are very much less; and I hardly like even to venture on the conjecture that the various "Nashes" along the coast are corruptions of *næs*. Here, in the Forest, we have no Danish "thorpes" or "bys." There are no Carlbys, as in Lincolnshire and Yorkshire, but plenty of Old-English Charltons, and Charlmoors, and Charlmeads. No Norse "forces" run here, as in the north of England, but only "rides." No "denes" open out to the sea, as in Durham, but only "chines" and "bunnies." No Jutish "ings" are dwelt in, as in Kent, but only "tons" and "leys." Here, in fact, the people of Cerdic have identified themselves with the land, and have left their impress, now unchanged for more than thirteen centuries, on all the towns, and hamlets, and homesteads.

Thus we find on the eastern side of the Forest, formerly in it, Eling, the Mark of the Ealingas, Totton, the Mark of the Totingas; on the south, Lymington and Pennington, the Marks of the Limingas and Penningas; and on the west, Fordingbridge, the Mark of the Fordingas, and Ellingham, that is, Adeling's Hamlet, Adelingeham, in *Domesday*, where some of the neighbouring woods are to this hour called Adlem's Plantations.

We will not press, as a proof of descent, the number of Old-English surnames, which may easily be collected in the district. We must remember that they were not used, in our modern sense, till long after the Norman Conquest; and when adopted, people were more likely to choose them from English, than from Norman or other sources. Such evidence establishes nothing.

In other ways, however, do we find the Old-English nomenclature telling us the history of the people and the country;—in Hengistbury Head, on the south-west, reminding us of the white horse—the Hengest of the High-German,[5] and Calshot at the east, spelt as we know in Edward I.'s time, Kalkesore; on the north-west in Charford—the old Cerdices-ford of *The Chronicle*; on the south in Darrat (Danes-rout) and Danestream, whose waters, the peasant maintains, still run red with the blood of the conquered.

Everywhere we meet similar compounds,—in Needshore, which the Ordnance map spells Needs-oar, and thus loses the etymology, which, like the Needle Rocks, means simply the under (German *nieder*) shore; in the various Galley Hills, corrupted into Gallows Hills, which have nothing to do with the later but the older instrument, which contained the signal-fires, and are connected with the words "galley," to frighten, and "galleybaggar," a scarecrow, still heard every day, from the Old-English *gælan*.[6]

We find the same impress in Lyndhurst, Brockenhurst, Ashurst, and, as we have before said, in various other hursts,[7] in the different Holmsleys, Netleys,[8] Beckleys, Bentleys, Bratleys, Stockleys, all from the Old-English *leag*; in the various *tons*, as Wootton, Winkton, Everton, Burton, and Hinton; in Gore and Goreley, the muddy places; in Culverley, the dove lea; in the Roydons and Rowdouns, the rough places; in Rhinefield, the hrook field; and in Brockis Hill, the hadger's hill.

Take only the very names of the fields and we shall meet the same element, as in the Wareham field, the fishing place field; Conygers and Coneygar,[9] the King's ground, to be met in every village; in the linches, as Goreley Linch, that is, Goreley Headland, or literally, the dirty-field headland; in Hangerley, the corner meadow; Hayes, the enclosure, with all its compounds, as Westhayes, Powelhayes, Crithayes, and Felthayes; in such terms as Withy Eyot, that is, Withy Island; and the different Rodfords—"hryðeranford"—the cattleford, the Old-English equivalent to the Norman Bovreford.

We meet, too, in daily life, such words as hayward for the North-Country "pinder;" barton, literally the barley place, instead of the Keltic "croo yard;"—the same Old-English element in the names of the flowers, as bishop-wort (*bisceop-wyrt*), one of the mints, from which the peasant makes his "hum-water;" cassock (from *cassuc*), any kind of binding weed, and cammock (from *cammec*), any of the St. John's worts, or ragworts; clivers (from *clife*, a bur), the heriff; and wythwind, by which name the convolvulus is still known, the Old-English "wið-winde."

These words, however, belong more especially to the next chapter. To descend from generals to particulars, let us notice some of the verbal characteristics by which a West-Saxon population may be distinguished. As a rule it may be laid down that the West-Saxons give a soft, and the Anglians and Northmen a hard sound to all their words. Thus in the New Forest we find the West-Saxons saying burrow for barrow; haish for harsh; pleu for plough; heth for heath, instead of the "hawth" of the Eastern rapes of Sussex; mash for marsh; Gerge for George; slue for sloe, and again, for slough, the "slow" of the north; bin for been, and also being; justle for jostle, as in Nahum, ch. ii. v. 4; athert for athwart; wool for hole; ballat for ballad, or, as it is pronounced in the more northern counties, ballard; ell for eel; clot and clit for clod; stiffle for stifle; ruff for roof, and so on. Thus, too, we meet here not with Deepdene, but Dibden, spelt in Boazio's map of 1591, Debden. No Chawton, but only Chewton occurs, no Farnham, but only Fernham and Fernhill.

The West-Saxons, too, have a peculiar drawl. So in the New Forest we may hear them saying pearts for parts; stwoane for stone; twereable for terrible; measter (*mæster*), instead of the Anglian "muster;" and yees instead of the Sussex "yus." As others have also remarked, the West Saxon substitutes *a* for *o*. So here we get lard for lord; nat for not; amang for among; knap for knop; shart for short; starm for storm; and Narmanton for Normanton. Not only this, but the West-Saxon in the New Forest substitutes *a* for *e*, as in agg for egg, and lag for leg. He not only retains the hard g, but gives a *k* when he can, as in kiver for cover, and aker for acorn, the "aitchorn" of the Anglian districts. Let us notice, too, that he always changes the *f* into a *v*, as vern for fern, vire for fire, evvets for effets, voam for foam, as written by Chaucer, vail for fall, and fitches for vetches, as we find it in Ezekiel, ch. iv. v. 9.

To go further into these distinctions is here impossible. As are the people so is the language. By an analysis of the published glossaries of Dorsetshire, Wiltshire, and Sussex, I find that the New Forest possesses above two-thirds of the two former, differing here and there only in pronunciation, whilst of the latter it scarcely possesses one-tenth, proving plainly that the people are West-Saxons rather than South, descendants of Cerdic more than of Ella.[10]

Turning from these minor characteristics, and looking at the people themselves as they once were, and as they now stand, much might be added as to the unequal race which the West-Saxon has run with the Anglian and the Northman, and its effect on his character. The most casual observer even, in going over so small a space as the New Forest, must have noticed how Nature has favoured the Northern and Midland counties in their sources of wealth and industry. The great home-trade of the Middle Ages has entirely deserted the South. Once, too, all our men-of-war sailed from what are now small ports on the south coast. Our fleets were manned by crews from the Isle of Wight, and Lymington, and Lyme, and the neighbouring harbours. The seamanship of the West-Country was England's right arm.[11] But now the ironfields of Staffordshire have put out the furnaces. The coal mines of Durham have destroyed the charcoal trade, and taken away the seamanship. The brine-pits of Cheshire have dried up the salterns which covered the south-western shores. Of course, this loss of material prosperity has told on the intelligence and morals of the district.

In the New Forest itself, till within the last thirty years, smuggling was a recognized calling. Lawlessness was the rule during the last century. Warner says that he had then seen twenty or thirty waggons laden with kegs, guarded by two or three hundred horsemen, each bearing two or three "tubs," coming over Hengistbury Head, making their way, in the open day, past Christchurch to the Forest. At Lymington, a troop of bandits took possession of the well-known Ambrose Cave, on the borders of the Forest, and carried on, not only smuggling, but wholesale burglary. The whole country was plundered. The soldiers were at last called out, the men tracked, and the cave entered. Booty to an enormous extent was found. The captain turned King's evidence, and confessed that he had murdered upwards of thirty people, whose bodies had been thrown down a well, where they were found.[12]

Such was the state of the New Forest in the last century. But as recently as thirty or forty years ago every labourer was either a poacher or a smuggler, very often a combination of the two. Boats were built from the Forest timber in many a barn; and to this day various fields far inland are still called "the dockyard mead." Crews of Foresters, armed with "swingels," such as the West-Saxons of Somerset fought with in the battle of Sedgemoor, defied the coastguard. The principal "runs" were made at Beckton and Chewton Bunnies, and the Gangway. Often as many as a hundred "tubs," each containing four gallons, and worth two or three guineas, or even more, would be run in a night. Each man would carry two or three of these kegs, one slung in front and two behind; or if the cliff was very steep, a chain of men was formed, and the tubs passed from hand to hand.

All this has been done within the memory of people not so very old. Men were killed at Milton. Old Becton Bunny House was burnt to the ground. A keg was carelessly broached, and the spirit caught fire from the spark of a pipe. Every person was in fact engaged in smuggling:—some for profit, many merely from a love of adventure. Everywhere was understood the smuggler's local proverb, "Keystone under the hearth, keystone under the horse's belly."[13]

Now nothing either in smuggling or poaching is to any extent attempted. In the one case the crime is unprofitable, in the other the temptation is withdrawn. Labour, too, is more plentiful, and the Government works of draining and planting in the Forest employ most of the Foresters.

Many a man, however, can still tell how he has baited a hook, tied to a bough, with apples to snare the deer; how he has pared the faun's hoof to keep the doe in one place, till he wanted to kill her. But now the deer are all gone, except a few, only seen now and then, wandering about in the wildest and loneliest parts. As to restocking the Forest, we can only say, with good Bishop Hoadley, respecting Waltham Chace,—"the deer have already done enough mischief."

The King's Gairn Brook

Footnotes

1. I may seem to exaggerate both here and in the next chapter. I wish that I did. For similar cases in the neighbouring counties of Dorset and Sussex let the reader turn to the words "hag-rod,"

"maiden-tree," and "viary-rings," in Mr. Barnes's *Glossary of the Dorset Dialect*; and vol. ii. pp. 266, 269, 270, 278, of Mr. Warter's *Seaboard and the Down*. I hesitate not to say that superstition in some sort or another is universal throughout England. It assumes different forms: in the higher classes, just at present, of spirit-rapping and table-turning, more gross than even those of the lower; and I am afraid really seems constitutional in our English nature.

2. Of the extreme difficulty of classification of race in the New Forest I am well aware. I have, however, taken such typical families as Purkis, Feckham, Watton, &c, whose names are to be met in every part of the Forest, as my guide. Often, too, certain Forest villages, as Burley and Minestead, though far apart, have a strong connection with each other, and a family relationship may be traced in all the cottages. A good paper was read, touching upon the elements of the New Forest population, by Mr. D. Mackintosh, before the Ethnological Society, April 3rd, 1861. Of the Jute element, which we might have expected from Bede's account of the large Jute settlement in the Isle of Wight, I find few traces. Still there are probably more than is generally supposed. See, however, on this point, what Latham says in his *Ethnology of the British Isles*, pp. 238, 239.

3. See Dr. Guest's paper on "The Early-English Settlements in South Britain," *Proceedings of the Archæological Institute*, Salisbury volume, 1851, p. 30.

4. This, of course, is not the place to go into so difficult a subject. I need not refer the reader to Mr. Davies's paper in the *Philological Society's Transactions*, 1855, p. 210, and M. de Haan Hettema's *Commentary* upon it, 1856, p. 196. On the great value of provincialisms, see what Müller has said in *The Science of Language*, pp. 49-59. In Appendix I., I have given a list of some of those of the New Forest, which have never before been noticed in any of the published glossaries.

5. In the charter of confirmation of Baldwin de Redvers to the Conventual House of Christchurch, quoted in Dugdale's *Monasticon Anglicanum*, vol. iii., part i., p. 304, and by Warner, vol. ii.. Appendix, p. 47, it is called Hedenes Buria, which may suggest that the word is only a corruption. I do not for one moment wish to insist on the personal reality of Hengest, but simply to notice the fact of the High-German word for a horse being prominent in the topography of a people whose ancestors used so many High-German words. See Donaldson, *Cambridge Essays*, 1856, pp. 45-48.
6. On this word see *Transactions of the Philological Society*, part i., 1858, pp. 123, 124.
7. See ch. iii., p. 33.
8. In the parish of Eling we have Netley Down and Netley Down-field, the Nutlei of *Domesday*. Upon this word—which we find, also, in the north of Hampshire, in the shape of Nately Scures and Upper Nately (Nataleie in *Domesday*)—as the equivalent of Natan Leah, the old name of the Upper portion of the New Forest, see Dr. Guest, as before quoted, p. 31.
9. A Keltic derivation has, I am aware, been proposed for this word. It is to be met with under various forms in all parts of the Forest. The Forest termination den (*denu*) must, however, be put down to this source. See *Transactions of the Philological Society*, 1855, p. 283.
10. See what Mr. Cooper says with regard to the affinity of the western dialect of Sussex, as distinguished from the eastern, to that of Hampshire, in the preface (p. i.) to his *Glossary of Provincialisms in the County of Sussex*. For instance, such Romance words as appleterre, gratten, ampery, bonker, common in Sussex, are not to be heard in the Forest; whilst many of the West-Country words, as they are called, used daily in the Forest, as charm (a noise—see next chapter, p 191), moot, stool, vinney, twiddle (to chirp), are, if Mr. Cooper's Glossary is correct, quite

unknown in Sussex.

11. It is surprising, in looking over the musters of ships in the reigns of Edward II. and Edward III., to see how few Northern ports are mentioned. The importance, too, of the South-coast ports, which were sometimes summoned by themselves, arose not only from the reasons in the text, but from being close to the country with which we were in a state of chronic warfare. See, too, the *State Papers*, vol. i., p. 812, 813, where the levies of the fleets in 1545, against D'Annebault, with the names of each vessel and its port, are given; as also p. 827, where the neighbouring coast of Dorset is described as deserted, in consequence of the sailors flocking to the King's service. I think that I have somewhere seen that our sailors were once rated as English, Irish, Scotch, and the "West Country," the latter standing the highest.

12. From an old chap-book, *The Hampshire Murderers*, with illustrations, without date or publisher's name, but probably written about 1776.

13 That is to say, the smuggled spirits were concealed either below the fireplace or in the stable, just beneath where the horse stood. The expression of "Hampshire and Wiltshire moon-rakers" had its origin in the Wiltshire peasants fishing up the contraband goods at night, brought through the Forest, and hid in the various ponds.

CHAPTER XVI.

THE FOLK-LORE AND PROVINCIALISMS.

Anderwood Corner

Intimately bound up with the race are of course the folk-lore of a district, and what we are now pleased to call provincialisms, but which are more properly nationalisms, showing us the real texture of our language; and in every way preferable to the Latin and Greek hybridisms, which are daily coined to suit the exigencies of commerce or science.

Provincialisms are, in fact, when properly looked at, not so much portions of the original foundations of a language, as the very quarry out of which it is hewn. And as if to compensate for much of the harm she has done, America has wrought one great good in preserving many a pregnant Old-English word, which we have been foolish enough to disown.[1] Provincialisms should be far more studied than they are; for they will help us to settle many a difficult point,—where was the boundary of the Anglian and the Frisian; how far on the national character was the influence of the Dane felt? how much, and in what way, did the Norman affect the daily business of life?

Still more important is a country's folk-lore, as showing the higher mental faculties of the race, in those legends and snatches of song, and fragments of popular poetry, which speak the popular feeling, and which not only contain its past history, but foreshadow the future literature of a country; in those proverbs, too, which tell the life and employment of a nation; and those superstitions which give us such an insight into its moral state.

Throughout the West of England still linger some few stray waifs and legends of the past. In the New Forest Sir Bevis of Southampton is no mythical personage, and the peasant will tell how the Knight used to take his afternoon's walk, across the Solent, from Leap to the Island.

Here in the Forest still dwell fairies. The mischievous sprite, Laurence, still holds men by his spell and makes them idle. If a peasant is lazy, it is proverbially said, "Laurence has got upon him," or, "He has got a touch of Laurence." He is still regarded with awe, and barrows are called after him. Here, too, in the Forest still lives Shakspeare's Puck, a veritable being, who causes the Forest colts to stray, carrying out word for word Shakspeare's description,—

"I am that merry wanderer of the night,
When I a fat and bean-fed horse beguile,
Neighing in likeness of a filly foal."

(Midsummer Night's Dream, Act ii., Sc. 1.)

This tricksy fairy, so the Forest peasant to this hour firmly believes, inhabits the bogs, and draws people into them, making merry, and laughing at their misfortunes, fulfilling his own roundelay—

"Up and down, up and down,
I will lead them up and down;
I am feared in field and town,
Goblin, lead them up and down."

(*Midsummer Night's Dream*, Act iv., Sc. 2.)

Only those who are eldest born are exempt from his spell. The proverb of "as ragged as a colt Pixey" is everywhere to be heard, and at which Drayton seems to hint in his *Court of Faerie*:—

"This Puck seems but a dreaming dolt,
Still walking like a ragged colt."

He does not, however, in the Forest, so much skim the milk, or play pranks with the chairs, but, as might be expected from the nature of the country, misleads people on the moors, turning himself into all sorts of shapes, as Shakspeare, Spenser, and Jonson, have sung. There is scarcely a village or hamlet in the Forest district which has not its "Pixey Field," and "Pixey Mead," or its "Picksmoor," and "Cold Pixey," and "Puck Piece." At Prior's Acre we find Puck's Hill, and not far from it lies the great wood of

Puckpits; whilst a large barrow on Beaulieu Common is known as the Pixey's Cave.[2]

Then, too, on the south-west borders of the Forest remains the legend, its inner meaning now perhaps forgotten, that the Priory Church of Christchurch was originally to have been built on the lonely St. Catherine's Hill, instead of in the valley where the people lived and needed religion. The stones, however, which were taken up the hill in the day were brought down in the night by unseen hands. The beams, too, which were found too short on the heights, were more than long enough in the town. The legend further runs, beautiful in its right interpretation, that when the building was going on, there was always one more workman—namely, Christ—than came on the pay-night.

So, too, the poetry of the district has its own characteristics, which it shares with that of the neighbouring western counties. The homeliness of the songs in the West of England strangely contrasts with the wild spirit of those of the North, founded as the latter so often are on the border forays and raids of former times. None which I have collected are direct enough in their bearing on the New Forest to warrant quotation, and I must content myself with this general expression.[3]

To pass on to other matters, let us notice some of the superstitions of the New Forest. No one is now so superstitious, because no one is so ignorant as the West-Saxon. One of the commonest remedies for consumption in the Forest is the "lungs of oak," a lichen (*Sticta pulmonaria*) which grows rather plentifully on the oak trees; and it is no unfrequent occurrence for a poor person to ask at a chemist's shop for a "pennyworth of lungs of oak." So, too, for weak eyes, "brighten," another lichen, is recommended. I do not know, however, that we must find so much fault in this matter, as the lichens were not very long ago favourite prescriptions with even medical men.

Again, another remedy for various diseases used to be the

scrapings from Sir John Chydioke's alabaster figure, in the Priory Church of Christchurch, which has, in consequence, been sadly injured. A specific, however, for consumption is still to kill a jay and place it in the embers till calcined, when it is then drunk at stated times in water. Hares' brains are recommended for infants prematurely born. Children suffering from fits are, or rather were, passed through cloven ash-trees. Bread baked on Good Friday will not only keep seven years, but is a remedy for certain complaints. The seventh son of a seventh son can perform cures. In fact, a pharmacopoeia of such superstitions might be compiled.

The New Forest peasant puts absolute faith in all traditions, believing as firmly in St. Swithin as his forefathers did when the saint was Bishop of Winchester; turns his money, if he has any, when he sees the new moon; fancies that a burn is a charm against leaving the house; that witches cannot cross over a brook; that the death's-head moth was only first seen after the execution of Charles I.; that the man in the moon was sent there for stealing wood from the Forest—a superstition, by the way, mentioned in a slightly different form by Reginald Pecock, Bishop of Chichester, in the fifteenth century.[4] And the "stolen bush," referred to by Caliban in the *Tempest* (Act ii., sc. 2), and Bottom in the *Midsummer Night's Dream* (Act vi. sc. 1), is still here called the "nitch," or bundle of faggots.[5]

Not only this, but the barrows on the plains are named after the fairies, and the peasant imagines, like the treasure-seekers of the Middle-Ages, that they contain untold wealth, and that the Forest wells are full of gold.[6]

I do not mean, however, to say that these beliefs are openly avowed, or will even be acknowledged by the first labourer who may be seen. The English peasant is at all times excessively chary—no one perhaps more so—of expressing his full mind; and a long time is required before a stranger can, if ever, gain his confidence. But I do say that these superstitions are all, with more or less credit, held in different parts of the Forest, although even many who believe them the firmest would shrink, from fear of ridicule, to confess the fact. Education has done something to remove them; but they have too firm a hold to be easily uprooted. They may not be openly expressed, but they are, for all that, to my certain knowledge, still latent.

Old customs and ceremonies still linger. Mummers still perform at Christmas. Old women "go gooding," as in other parts of England, on St. Thomas's Day. Boys and girls "go shroving" on Ash Wednesday; that is, begging for meat and drink at the farm-houses, singing this rude snatch:—

"I come a shroving, a shroving,
For a piece of pancake,
For a piece of truffle-cheese[7]
Of your own making."

When, if nothing is given, they throw stones and shards at the door.[8]

Plenty, too, of old love superstitions remain—about ash boughs with an even number of leaves, and "four-leaved" clover, concerning which runs a Forest rhyme:—

"Even ash and four-leaved clover,
You are sure your love to see
Before the day is over."

Then, too, we must not forget the Forest proverbs. "Wood Fidley rain," "Hampshire and Wiltshire moonrakers," and "Keystone under the hearth," have already been noticed. But there are others such as "As yellow as a kite's claw," "An iron windfall," for anything unfairly taken, "All in a copse," that is, indistinct, "A good bark-year makes a good wheat-year," and "Like a swarm of bees all in a charm," explained further on, which show the nature of the country. Again, "A poor dry thing, let it go," a sort of poacher's euphemism, like, "The grapes are sour," is said of the Forest hares when the dogs cannot catch them, and so applied to things which are coveted but out of reach. "As bad as Jeffreys" preserves, as throughout the West of England, the memory of one who, instead of being the judge, should have been the hangman. Again, too, "Eat your own side, speckle-back," is a common Forest expression, and is used in reference to greedy people. It is said to have taken its origin from a girl who shared her breakfast with a snake, and thus reproved her favourite when he took too much. Again, "To rattle like a boar in a holme bush," is a thorough proverb of the Forest district, where a "holme" bush means an old holly. Passing, however, from particulars to generals, let me add for the last, "There is but one good mother-in-law, and she is dead." I have never heard it elsewhere in England, but doubtless it is common enough. It exactly corresponds with the German saying, "There is no good mother-in-law but she that wears a green gown," that is, who lies in the churchyard. The shrewdness and humour of a people are never better seen than in their proverbs.

Further, there are plenty of local sayings, such as "The cuckoo goes to Beaulieu Fair to buy him a greatcoat," referring to the arrival of the cuckoo about the 15th of April, whilst the day on which the fair is held is known as the "cuckoo day." A similar proverb is to be found in nearly every county. So, also, the saying with regard to Burley and its crop of mast and acorns may be met in the Midland districts concerning Pershore and its cherries. Like all other parts of England, the Forest is full, too, of those sayings and adages, which are constantly in the mouths of the lower classes, so remarkable for their combination of both terseness and metaphor. To give an instance, "He won't climb up May Hill," that is, he will not live through the cold spring. Again, "A dog is made fat in two meals," is applied to upstart or purse-proud people. But it is dangerous to assign them to any particular district, as by their applicability they have spread far and wide.

One or two historical traditions, too, still linger in the Forest, but their value we have seen with regard to the death of the Red King. Thus, the peasant will tell of the French fleet, which, in June, 1690, lay off the Needles, and of the Battle of Beachy Head, whose cannonading was heard even in the Forest, but who fought, or why, he is equally ignorant. One tradition, however, ought to be told concerning the terrible winter of 1787, still known in the Forest as "the hard year." My informant, an old man, derived his knowledge from his father, who lived in the Forest in a small lonely farm-house. The storm began in the night; and when his father rose in the morning he could not, on account of the snow-drift, open the door. Luckily, a back room had been converted into a fuel-house, and his wife had laid in a stock of provisions. The storm still increased. The straggling hedges were soon covered; and by-and-by the woods themselves disappeared. After a week's snow, a heavy frost followed. The snow hardened. People went out shooting, and wherever a breathing-hole in the snow appeared, fired, and nearly always killed a hare.[9] The snow continued on the ground for seven weeks; and when it melted, the stiffened bodies of horses and deer covered the plains.[10]

And now for a few of the Forest words and expressions, many of which are very peculiar. Take, for instance, the term "shade," which here has nothing in common with the shadows of the woods, but means either a pool or an open piece of ground, generally on a hill top, where the cattle in the warm weather collect, or, as the phrase is, "come to shade," for the sake of the water in the one and the breeze in the other. Thus "Ober Shade" means nothing more than Ober pond; whilst "Stony Cross

Shade" is a mere turfy plot. At times as many as a hundred cows or horses are collected together in one of these places, where the owners, or "Forest marksmen," always first go to look after a strayed animal. Nearly every "Walk" in the Forest has its own "Shade," called after its own name, and we find the term used as far back as a perambulation of the Forest in the twenty-second year of Charles II., where is mentioned "the Green Shade of Biericombe or Bircombe."

It affords a good illustration of how words grow in their meaning, and imperceptibly pass from one stage to another. It originally signified nothing but a shadow, and then the place where the shadow rests. In this second meaning it more particularly became associated with the idea of coolness, but gradually, whilst acquiring that idea, quite contrary to Milton's "unpierced shade" (*Paradise Lost*, B. iv. 245), lost the notion of that coolness being caused by the interception of light and heat, and was thus transferred to any place which was cool, and so at last applied, as in the New Forest, to bare spots without a tree, deriving their coolness either from the breeze or the water.

Another instance of the gradual change in the meaning of words amongst provincialisms may be found in "scale," or "squoyle." In the New Forest it properly signifies a short stick loaded at one end with lead, answering to the "libbet" of Sussex, and is distinguished from a "snog," which is only weighted with wood. With it also is employed the verb "to squoyle," better known in reference to the old sport of "cock-squoyling." From throwing at the squirrel the word was used in reference to persons, so that "Don't squoyle at me" at length meant, "Do not slander me." Lastly, the phrase, now still common, "Don't throw squoyles at me," comes by that forced interpretation of obtaining a sense, which nearly always reverses the original meaning, to signify, "Do not throw glances at me." And so in the New Forest at this day "squoyles" not unfrequently mean glances.

There is, too, the word "hat," which in the Forest takes the place of "clump," and is nearly equivalent to the Sussex expression, "a toll of trees." I have no doubt whatever that the word had its origin in the high-crowned hats of the Puritans, the "long crown" of the proverb; and in the first place referred only to tall isolated clumps of trees. Now, however, it does not merely mean a clump or ring, as the "seven firs" between Burley and Ringwood, and Birchen, and Dark Hats, near Lyndhurst, but any small irregular mass of trees, as the Withy Bed Hat in the valley near Boldrewood.

Then of course, in connection with the Forest trees, many peculiar words occur. The flower of the oak is called "the trail," and the oak-apple the "sheets axe,"—children carrying it on the twenty-ninth of May, and calling out the word in derision to those who are not so provided. The mast and acorns are collectively known as "the turn out," or "ovest;"[11] whilst the badly-grown or stunted trees are called "bustle-headed," equivalent to the "oak-barrens" of America.

Other words there are, too, all proclaiming the woody nature of the country. The tops of the oaks are termed, when lopped, the "flitterings," corresponding to the "batlins" of Suffolk. The brush-wood is still occasionally Chaucer's "rise," or "rice," connected with the German reis; and the beam tree, on account of its silvery leaves, the "white rice."[12] Frith, too, still means copse-wood. The stem of the ivy is the "ivy-drum." Stumps of trees are known as "stools," and a "stooled stick" is used in opposition to "maiden timber," which has never been touched with the axe; whilst the roots are called "mocks," "mootes," "motes," and "mores." But about these last, which are all used with nice shades of difference, we shall have, further on, something to say.

Nor must we forget the bees which are largely kept throughout the Forest, feeding on the heather, leading Fuller to remark that Hampshire produced the best and worst honey in England.

The bee-season, as it is called, generally lasts, on account of the heath, a month longer than on the Wiltshire downs. A great quantity of the Old-English mead—*medu*—is still made, and it is sold at much the same value as with the Old-English, being three or four times the price of common beer, with which it is often drunk. The bees, in fact, still maintain an important place in the popular local bye-laws. Even in *Domesday* the woods round Eling are mentioned as yearly yielding twelve pounds' weight of honey. As may therefore be expected, when we remember that the whole of England was once called the Honey Island, here, as elsewhere, plenty of provincialisms occur concerning the bees.[13]

The drones are here named "the big bees," the former word being in some parts seldom used. The young are never said to swarm, but "to play," the word taking its origin from their peculiar flight at the time: as Patmore writes,—

"Under the chestnuts new bees are swarming,
Falling and rising like magical smoke."

The caps of straw which are placed over the "bee-pots," to protect them from wet, are known as the "bee-hackles," or "bee-hakes." This is one of those expressive words which is now only found in this form, and that, in the Midland Counties, of "wheat hackling," that is, covering the sheaves with others in a peculiar way, to shelter them from the rain. About the honeycombs, or, as they are more commonly called, "workings," the following rhyme exists:—

"Sieve upon herder,[14]
One upon the other;
Holes upon both sides,
Not all the way, though,
What may it be? See if you know."

The entrance for the bees into the hive is here, as in Cambridgeshire and some other counties, named the "tee-hole," evidently an onomatopoieia, from the buzzing or "teeing" noise, as it is locally called, which the bees make. The piece of wood placed under the "bee-pots," to give the bees more room, is known as "the rear," still also, I believe, in use in America. The old superstition, I may notice, is here more or less believed, that the bees must be told if any death happens in a family, or they will desert their hives. It is held, too, rather, perhaps, as a tradition than a law, that if a swarm of bees flies away the owner cannot claim them, unless, at the time, he has made a noise with a kettle or tongs to give his neighbours notice. It is on such occasions that the phrase "Low brown" may be heard, meaning that the bees, or the "brownies," as they are called, are to settle low.

So also of the cattle, which are turned out in the Forest, we find some curious expressions. A "shadow cow" is here what would in other places be called "sheeted," or "saddle-backed," that is, a cow whose body is a different colour to its hind and fore parts.[15] A "huff" of cattle means a drove or herd, whilst the cattle, which are entered in the marksman's books, are said to he "wood-roughed." A cow without horns is still called a "not cow" (*hnot*), exactly corresponding to the American "humble" or "bumble cow," that is, shorn, illustrating, as Mr. Akerman notices,[16] Chaucer's line,—

> "A not hed had he, with a brown visage."

In the Forest, too, as in all other districts, a noticeable point is the number of words formed by the process of onomatopoieia. Thus, to take a few examples, we have the expressive verb to "scroop," meaning to creak, or grate, as a door does on rusty hinges; and again the word "hooi," applied to the wind whistling round a corner, or through the key-hole, making the sound correspond to the sense. It exactly represents the harsh creaking, as the Latin *susurrus* and the Ψιθύρισμα of Theocritus reflect the whisperings of the wind in the pines and poplars, resembling, as Tennyson says, "a noise of falling showers." Again, such words as "clocking," "gloxing," applied to falling, gurgling water; "grizing," and "snaggling," said of a dog snarling; "whittering," or "whickering"—exactly equivalent to the German *wichern*—of a young colt's neighing; "belloking," of a cow's lowing—are all here commonly used, and are similarly formed. Names of animals take their origin in the same way. The wry-neck, called the "barley-bird" in Wiltshire, and the "cuckoo's mate" and "messenger" elsewhere, is in the Forest known as the "weet-bird," from its peculiar cry of "weet," which it will repeat at short intervals for an hour together. So, too, the common green woodpecker is here, as in some other parts of England, called, from its loud shrill laugh, the "yaffingale." The goat-sucker, too, is the "jar-bird," so known from its jarring noise, which has made the Welsh name it the "wheelbird" (*aderyn y droell*), and the Warwickshire peasant the "spinning-jenny." In fact, a large number of birds in every language are thus called, and to this day in the cry of the peacock we may plainly hear its Greek name, ταῶς.

Of course, I need not say, we must be on our guard against adopting the onomatopoetic theory as explaining the origin of language. Within, however, certain limits, especially with a peculiar class of provincialisms, it gives us, as here, some aid.[17]

Again, as an example of phrases used by our Elizabethan poets, preserved only by our peasantry, though in good use in America, take the word "bottom," so common throughout the Forest, meaning a valley, glen, or glade. Beaumont and Fletcher and Shakspeare frequently employ it. Even Milton, in *Paradise Regained*, says—

"But cottage, herd, or sheepcote, none he saw,
Only in a bottom saw a pleasant grove."

(Book ii. 289.)

In his *Comus*, too, we find him using the compound "bottomglade," just as the Americans speak to this day of the "bottomlands" of the Ohio, and our own peasants of Slufter Bottom, and Longslade Bottom, in the New Forest.

"Heft," too, is another similar instance of an Old-English word in good use in America and to be found in the best American authors, but here in England only employed by our rustics. To "heft" (from *hebban*, with the inflexions, *hefest*, "hefð," still used), signifies to lift, with the implied meaning of weighing. So, "to heft the bee-pots," is to lift them in order to feel how much honey they contain. The substantive "heft" is used for weight, as, "the heft of the branches."

Again, also, the good Old-English word "loute" (*lutan*), to bend, bow, and so to touch the hat, to be heard everyday in the Forest, though nearly forgotten elsewhere in England, may be found in Longfellow's *Children of the Lord's Supper*:—

"as oft as they named the Redeemer,
Lowly louted the boys, and lowly the maidens all
courtesied."

In fact, one-half of the words which are considered Americanisms are good Old-English words, which we have been foolish enough to discard.

Let us now take another class of words, which will help to explain difficult or corrupt passages in our poets. There is, for instance, the word "bugle" (*buculus*), meaning an ox (used, as Mr. Wedgwood[18] notices, in Deut. xiv. in the Bible, 1551), which is forgotten even by the peasantry, and only to be seen, as at Lymington and elsewhere, on a few inn-signs, with a picture sometimes of a cow, by way of explanation. I have more than once thought, when Rosalind, in *As You Like it* (Act iii., sc. 5), speaks of Phoebe's "bugle eyeballs," she means not merely her sparkling eyes, as the notes say, but rather her large, expressive eyes, in the sense in which Homer calls Minerva βοῶπις.

To give one more illustration of the value of provincialisms in such cases, let us take the word "bumble," which not only in the New Forest means, in its onomatopoetic sense, to buzz, hum, or boom, as in the common proverb, "to bumble like a bee in a tar-tub," and as Chaucer says, in *The Wife of Bath's Tale*—

"A bytoure bumbleth in the myre,"

but is also used of people stumbling or halting. Probably, in *The Merry Wives of Windsor* (Act iii., sc. 8), in the passage which has been of such difficulty to the commentators, where Mrs. Ford says to the servants, who are carrying Falstaffe in the buck-basket—"Look, how you drumble," which has no meaning at all, we should, instead, read this word. It, at all events, not only conveys good sense, but is the exact kind of word which the passage seems to expect.

Again, the compound thiller-horse, from the Old-English "bill," a beam or shaft, and so, literally, the shaft-horse, which we find in Shakspeare under the form of "thill-horse" (*Merchant of Venice*, Act ii., sc. 2), is here commonly used.

Then there are other forms among provincialisms which give such an insight into the formation of language, and show the common mind of the human race. Thus, take the word "three-cunning,"[19] to be heard every day in the Forest, where three has the signification of intensity, just as the Greek τρίς in composition in the compounds τρίσμακαρ, τρίσάθλιος, and other forms. So, too, the missel-thrush is called the "bull-thrush," with the meaning of size attached to the word, as it does more commonly to our own "horse," and the Greek ἵππος, and the Old-English *hrefen*, raven, in composition.

As might be expected, from what we have seen of the population of the Forest, the Romance element in its provincialisms is very small. Some few words, such as "merry," for a cherry; "fogey," for passionate; "futy," for foolish; "rue," for a hedge; "glutch," to stifle a sob—have crept in, besides such Forest terms as verderer, regarder, agister, agistment, &c, but the majority are Teutonic. Old-English inflexions, too, still remain. Such plurals as placen, housen, peasen, gripen, fuzzen, ashen, hosen, as we find the word in Daniel, ch. iii. v. 21; such perfects as crope, from creep; lod, from lead; fotch, from fetch; and such phrases as "thissum" ("þissum"), and "thic" for that, are daily to be heard.

Let us, for instance, take the adjective vinney, evidently from the Old-English *finie*, signifying, in the first place, mouldy; and since mould is generally blue or purplish, it had gradually attached to it the signification of colour. Thus we find the mouldy cheese not only named "vinney," but a roan heifer called a "vinney heifer." The most singular part, however, as exemplifying the changes of words, remains to be told. Since cheese, from its colour, was called "vinney," the word was applied to some particular cheese, which was mouldier and bluer than others, and the adjective was thus changed into a substantive. And we now have "vinney," and the tautology, "blue vinney," as the names of a particular kind of cheese as distinguished from the other local cheeses, known as "ommary" and "rammel."[20]

So also with the word "charm," or rather "churm," signifying, in the first place, noise or disturbance, from the Old-English *cyrm*. We meet it every day in the common Forest proverb, "Like a swarm of bees all in a churm," whilst the fowlers on the coast talk also of the wild ducks "being in a churm," when they are in confusion, flapping their wings before they settle or rise. We find it, too, in the old Wiltshire song of the "Owl's Mishap," to be sometimes heard on the northern borders of the Forest:—

"At last a hunted zo ver away,
That the zun kum peping auver the hills,
And the burds wakin up they did un espy,
And wur arl in a churm az un whetted their bills."

The word was doubtless in the first place an onomatopoieia, denoting the humming, buzzing sound of wings. Since, however, it was particularly connected with birds, it seems to have been used in the sense of music and song by our Elizabethan poets, and by Milton. Thus:—

"Sweet is the breath of morn, her rising sweet
With charm of earliest birds."

(*Paradise Lost*, Book iv. 642.)

And again:—

"Morn when she ascends
With charm of earliest birds."

(*Paradise Lost*, Book iv. 651.)

Here, however, in the New Forest, we find the original signification of the word preserved.

Let us further notice one or two more words, which are used by Milton and his contemporaries, and even much later, but which are now found in the Forest, and doubtless elsewhere, as mere provincialisms. Thus, though we do not meet his "tale," in the sense of number, as in *L'Allegro*—

"And every shepherd tells his tale,
Under the hawthorn in the dale;"

—that is, number of sheep: we find its allied word "toll," to count. "I toll ten cows," is no very uncommon expression. Then, too, we have the word "tole," used, as I believe it still is in America, of enticing animals, and thus metaphorically applied to other matters. So, in this last sense, Milton speaks of the title of a book, "Hung out like a toling sign-post to call passengers."[21]

Again, too, the bat is here called "rere-mouse" (from the Old-English *hrere-mus*, literally the raw-mouse), with its varieties rennie-mouse and reiny-mouse,[22] whilst the adjective "rere" is sometimes used, as in Wiltshire, for raw. On the other hand, the word fliddermouse, or, as in the eastern division of Sussex, flindermouse (from the High-German *fledermaus*), does not, to my knowledge, occur. In the Midland counties it is often known as "leathern wings" (compare *ledermus*); and thus, Shakspeare, with his large vocabulary, using up every phrase and metaphor which he ever met, makes Titania say of her fairies:—

"Some war with rear-mice for their leathern wings."

(*Midsummer Night's Dream*, Act ii., sc. 3.)

To take a few words common, not only to the New Forest, but to various parts of the West of England, we shall see how strong is the Old-English element here in the common speech. The housewife still baits (*betan*, literally to repair, and so, when joined with *fyr*, to light) the fire, and on cold days makes it blissy (connected with *blysa*, a torch). The crow-boy in the spring sets up a gally-bagger (*gælan*, in its last meaning to terrify), instead of the "maukin" of the north, to frighten away the birds from the seed; and the shepherd still tends his chilver-lamb (*cilferlamb*) in the barton (*bere tun*, literally the barley enclosure). The labourer still sits under the lew (*hleow*, or "hleowð," shelter, warmth) of the hedge, which he has been ethering ("eðer," a hedge); and drives the stout (*stut*, a gadfly) away from his horses; and feels himself lear (*lærnes*, emptiness), before he eats his nammit (*nón-mete*), or his dew-bit (*deaw-bite*).

If we will only open our Bible we shall there find many an old word which could be better explained by the Forest peasants than any one else. Here the ploughman still talks of his "dredge," or rather "drudge," that is, oats mixed with barley, just as we find the word used in the marginal reading of Job xxiv. v. 6. Here, too, as in Amos (chap, iv., v. 9), and other places, the caterpillar is called the "palmer-worm." Here, also, as in other parts of England, the word "lease," from the Old-English *lesan*, is far commoner than glean, and is used just as we find it in Wycliffe's Bible, Lev. xix., 10:—" In thi vyneyeerd the reysonus and cornes fallynge down thou shalt not gedere, but to pore men and pilgrimes to ben lesid thou shalt leeve." The goatsucker is known, as we have seen, not only as the "jar-bird," but as the "night-hawk," as in Leviticus (chap, xi., v. 16) and Deuteronomy (chap, xiv., v. 15); and also the "night-crow," as we find it called in Barker's Bible (1616) in the same passages. So also the word "mote," in the well-known passage in St. Matthew (chap, viii., v. 3), is not here obsolete. The peasant in the Forest speaks of the "motes," that is, the stumps and roots of trees, in opposition to the smaller "mores," applied also to the fibres of ferns and furze, whilst the sailor on the coast calls the former "mootes," when he dredges them up in the Channel.[23]

With this I must stop. I will only add that the study of the West-Saxon dialect in the counties of Hants, Wilts, and Dorset, is all-important. As we go westward we shall find it less pure, and more mixed with Keltic. As is well known, the Britons lived with the Old-English in perfect harmony in Exeter. Their traces remain there to this day. In these three counties, therefore, are the most perfect specimens of the West-Saxon dialect to be found. Mr. Thorpe has noticed in the Old-English text of *Orosius*, which is now generally ascribed to Alfred, the change of *a* into *o* and *o* into *a*, and also the same peculiarity in Alfred's *Boethius*.[24] This we have already, in the last chapter, seen to be purely West-Saxon. I have no doubt whatever that at even the present day it is not too late to find other points of similarity, and make still clearer the West-Saxon origin of the Corpus Christi manuscript of the *Chronicle*,[25] and how far even Alfred and St. Swithin contributed to its pages. These are difficult questions; but I feel sure that much additional light can even yet be obtained. Sound criticism would show as much difference between our local dialects, whether even Anglian, or South, or West-Saxon, as between the Doric and Attic of Greece. I have dealt only with the broader features of the Old-English tongue, as it is still spoken in the Forest. Enough, however, I trust, has been shown of the value of provincialisms, even when collected over so limited a space. Everywhere in England we shall find Teutonic words, which are not so much the mould into which all other forms have been cast, as the living germ of our language. Mixed and imbedded with these, as we have also seen, we shall meet Keltic and Romance, by both of which our language has been so influenced and modified. Let us not be ashamed to collect them; for by them we may explain not only obscure passages in our old authors, but doubtful points in our very history.

Bushey Bratley (Another View).

Footnotes

1. See *Dictionary of Americanisms*, by J. E. Bartlett, who does not, however, we think, refer nearly often enough to the mother-country for the sources of many of the phrases and words which he gives. Even the Old-English inflexions, as he remarks, are in some parts of the States still used, showing what vitality, even when transplanted, there is in our language. Boucher, too, notices in the excellent introduction to his *Glossary of Archaic and Provincial Words*, p. ix., that the whine and the drawl of the first Puritan emigrants may still in places be detected.

2. All over the world lives a similar fairy, the same in form, but

different in name. His life has been well illustrated in Dr. Bell's *Shakspeare's Puck and his Folk-lore.* In England he is known by many names—"the white witch," "the horse-hag," and "Fairy Hob;" and hence, too, we here get Hob's Hill and Hob's Hole. For accounts of him in different parts see especially Allies' *Folk-lore of Worcestershire*, ch. xii. p. 409, and *Illustrations of the Fairy Mythology of A Midsummer Night's Dream*, by J. O. Halliwell. Published by the Shakspeare Society.

3. The most popular songs which I have noticed in the Forest and on its borders are the famous satire, "When Joan's ale was new," which differs in many important points from Mr. Bell's printed version: "King Arthur had three sons:" "There was an old miller of Devonshire," which also differs from Mr. Bell's copy; and

"There were three men came from the north,
To fight the victory;"

made famous by Burns' additions and improvements; but which, from various expressions, seems to have been, first of all, a West-Country song, sung at different wakes and fairs, part of the unwritten poetry of the nation.

4. *The Repression of Over-much Blaming the Church*, edited by Churchill Babington, vol. i., part, ii., ch. iii., p. 155.

5. Dr. Bell takes quite a different view of these passages in his *Shakspeare's Puck and his Folk-lore*. Introduction to vol. ii. p. 6. The simple explanation, however, seems to me the best.

6 See ch. xviii. p. 197.

7. The best cheese, the same as "rammel," as opposed to "ommary," which see in Appendix I.

8. In the Abstract of Forest Claims made in 1670 some old customs are preserved, amongst them payments of "Hocktide money," "moneth money," "wrather money" (rother, hryðer, cattle-money), "turfdele money," and "smoke money," which last we shall meet in the churchwardens' books of the district. The following is taken from the Bishop of Winchester's payments:—"

Rents at the feast of St. Michael, 3*s.* 8*d.* For turfdeale money, 3*s.* 0*d.* Three quarters and 4 bushels of barley at the feast of All Saints. Three bushels of oats, and 30 eggs, at the Purification of the Virgin Mary."—(p. 57.)

9. Against tracking hares on the snow and killing them with "dogge or beche bow," was one of the statutes of Henry VIII., made 1523 (*Statutes of the Realm*, vol. iii., p. 217).

10. In that winter 300 deer were starved to death in Boldrewood Walk. *Journals of the House of Commons*, vol. xliv., pp. 561, 594

11. I have never in the Forest met the old phrase of "shaketime," or rather "shack-time," as it should be written, and still used of the pigs going in companies after grain or acorns, according to Miss Gurney, in Norfolk. *Transactions of the Philological Society*, 1855, p. 35.

12. On this word, see Appendix I.

13. By a decree of the Court of Exchequer, in the twenty-sixth year of Elizabeth, the keepers were allowed to take all the honey found in the trees in the Forest.

14. A local name for a sieve, called, also, a "rudder;" which last word is, in different forms, used throughout the West of England.

15. For other words applied to cows of various colours, see Barnes's *Glossary of the Dorset Dialect*, under the words "capple-cow," p. 323; "hawked cow," p. 346; and "linded cow," p. 358.

16. Glossary of the Provincial Words and Places in Wiltshire, pp. 37, 38. London, 1842.

17. See Müller's *Science of Language*, pp. 345-351; and compare Wedgewood, *Dictionary of English Etymology*, introduction, pp. 5-17.

18. Dictionary of English Etymology, p. 260. Man wood uses "bugalles" as a translation of *buculi. A Treatise of the Lawes of the Forest*, f. iii., sect, xxvii., 1615.

19. Cunning, I need scarcely add, is here used in its original sense of knowing, from the Old-English *cunnan*, as we find in Psalm

cxxxvii. v. 5.
20. See ch. xvi. p. 178.
21. *Apology for Smectymnus*, quoted by Richardson. The word is even used by Locke.
22. Miss Gurney, in her *Glossary of Norfolk Words*, gives "ranny" as a shrew-mouse. *Transactions of the Philological Society*', 1855, p. 35. The change of *e* into *a* is worth noticing, as illustrative of what was said in the previous chapter, p. 167, of the pronunciation of the West-Saxon.
23. The word "more" was in good use less than a century ago; whilst the term "morefall," as we have seen in chapter iv. p. 43, foot-note, was very common in the time of the Stuarts. Mr. Barnes, in his *Glossary of the Dorset Dialect*, pp. 363, 391, gives us "mote," and "stramote," as "a stalk of grass," which serve still better to explain St. Matthew.
25. Thorpe's Preface to the English translation of Pauli's *Life of Alfred the Great*, p. vi.
25. Thorpe's Preface to *The Chronicle*, vol. i., p. viii., foot-note 1. See, however, Lappenberg's *History of England under the Anglo-Saxon Kings*; translated by Thorpe, Literary Introduction, p. xxxix.; and the Preface to *Monumenta Historica Britannica*, p. 75, where, as Mr. Thorpe notices, the examples quoted, in favour of the Mercian origin of the manuscript, are certainly, in several instances, wrong.

CHAPTER XVII.

THE BARROWS.

The Urns in Bratley Barrow.

IT is much to be regretted that Sir Walter Scott has left no account of his excavations of various barrows in the Forest. However little we may be able to determine by the evidence, or however conjectural the inferences which we may draw, there will, at least, be this value to this chapter, that it will put on record facts which otherwise could not be known.

The barrows lie scattered all over the Forest, and are known to the Foresters by the name of "butts," some of the largest being distinguished by local appellations. As in other parts of England, and as in France, superstition connects them with the fairies; and so we find on Beaulieu Plain two mounds known as the Pixey's Cave and Laurence's Barrow.

My own excavations have been entirely confined to the Keltic barrows in the northern part of the Forest.[1] But we will first of all take those on Sway and Shirley Commons, opened by Warner.[2] The largest stands a little to the east of Shirley Holms, close to Fetmoor Pond, measuring about a hundred yards in circumference, and surrounded by three smaller mounds varying from thirty to fifty yards, and two more nearly indistinct. These two last are, I suspect, those opened by Warner, where, after piercing the mound, he found on the natural soil a layer of burnt earth mixed with charcoal, and below this, at the depth of two feet, a small coarse urn with "an inverted brim,"[3] containing ashes and calcined bones.

Some more lie to the northward, and are distinguished by being trenched. Two of these also were opened by Warner, but he failed to discover anything beyond charcoal and burnt earth.

His opinion was that these last belonged to the West-Saxons and the former to the Kelts, who were slain defending their country against Cerdic. So large a generalization, however, requires far stronger evidence than can at present be produced.

Warner, too, is besides wrong in much of his criticism, such as that the Teutonic nations never practised urn-burial; whilst the banks in which he sees fortifications may be only the embankments within which dwelt a British population.

Still there is some probability about the conjecture. A little farther down the Brockenhurst stream are Ambrose Hole and Ampress Farm, both names unmistakeably referring to Ambrosius, or Natan-Leod, the Roman general who led the Britons against their invaders. Nearer Lymington, too, stands Buckland Rings,[4] a Roman camp, with its south and north sides still nearly perfect, to which, perhaps, Natan-Leod fell back from Calshot.

All this, however, must be accepted as mere conjecture. A more critical examination of these barrows is still wanting.

Close to them, however, lies Latchmoor or Lichmoor Pond, the moor of corpses, a name which we meet again a little to the westward in Latchmoor Water, which flows by Ashley Common. The words are noticeable, and in connection with Darrat's (Dane-rout) stream, which is also not far distant, may point to a very different invasion.[5]

And now we will pass to the barrows which I have opened. The first are situated on Bratley Plain, as the name shows, a wide heath, marked only by a few hollies and the undulations of the scattered mounds. The largest barrow lies close to the sixth milestone on the Ringwood Road. In a straight line to the north, at the distance of a quarter of a mile apart, rise three others, whilst round it on the east side lie a quantity of small circles, so low as hardly to be discernible when the heather is in bloom. An irregularly shaped oval, it rose in the centre to a height of nearly six feet above the ground, measuring sixteen yards in breadth, and twenty-two in length, with a circumference of from sixty to sixty-five. On the south side was a depression from whence the gravel had been obtained. We first cut a trench two yards broad, so as to take the centre, and at about two feet and a half from the surface came upon traces of charcoal, which increased till we reached the floor. A few round stones, probably, as they bore some slight artificial marks, used for slinging, and the flake of, perhaps, a flint knife, were the only things found, and were all placed on the south side. We now cut the mound from east to west, and on the east side, resting on the floor, we discovered the remains of a Keltic urn. The parts were, however, in a most fragile state, and in some instances had resolved themselves into mere clay, and we could only obtain two small fragments, sufficient to show the coarseness and extreme early age of the ware. No charcoal nor osseous matter could be detected adhering to the sides, which, as we shall see, is generally the case.

Round it, as was stated, lie a quantity of small grave-circles, varying from twenty-five to ten yards in circumference, and scarcely better defined than fairy-rings. Two of these I opened, and they corresponded with the mounds on Sway Common examined by Warner, in having a grave about three feet deep, in which we found only charcoal. This was, however, the only point of resemblance, as they had no mound, and contained no urn. One fact is worth noticing, that they were dug in a remarkably hard gravelly soil, so hard that the labourers made very slow progress even with their pick-axes. I did not excavate any more, as they were all evidently of the same character. The choice of such a soil, especially with the instruments they possessed, may, perhaps, show the importance which the Britons attached to the rite of burial.

About a quarter of a mile, or rather less, from this great graveyard lay a solitary mound, two feet and a half in height, having a circumference of twenty-seven feet, a very common measurement, but without any trench. Upon digging into it on the east side we quickly came, about four inches from the surface, upon a patch of charcoal and burnt earth. Proceeding farther, we reached two well-defined layers of charcoal, the uppermost two feet from the top of the barrow. A band of red burnt earth, measuring five inches, separated these two beds, in both of which in places appeared white spots and patches of limy matter, the remains of calcined bones. In the centre, as shown in the drawing, we found a Keltic urn. Imbedded in a fine white burnt clay, which had hardened, placed with its mouth uppermost, and ornamented with a rough cable-moulding, and two small ears, it stood on the level of the natural soil, rising to within sixteen inches of the top of the mound.

Digging on both sides, we discovered two more urns imbedded in the same hard white sandy clay, so hard that it had to be scraped away with knives. Like the first, they were made by hand, and when exposed quite shone with a bright vermilion, which quickly changed to a dull grey. The paste, however, was a light yellow, mixed with coarse gritty sand. And the three were placed, as shown by the compass, exactly due north-east and south-west.

A plain moulding ran round the south-west urn, which was considerably smaller and not so well baked as the other two, and had very much fallen to pieces from natural decay. This was placed eight inches lower than the central urn.

The northernmost was the same size as the central, though differing from it in the contraction of the rim, and when discovered was perfectly whole, but was unfortunately fractured by being separated from a large furze root, which had completely twined round the upper part. It, too, was placed on a lower level, by four inches, than the central urn. The two extreme urns were exactly five feet apart, and the interiors of them all were blackened by the carbon from the charcoal, burnt earth, and bones, which they contained.

Looking at their rude forms and large size, their straight sides, their wide mouths, the thickness, and the rough gritty texture of the paste,[6] the absence of nearly all ornamentation, and, with the exception, perhaps, of a slinging stone, of all weapons, we shall not be wrong in dating them as long anterior to the Roman invasion—how long a more minute criticism and a greater accumulation of facts than is now possessed, can alone determine.

There are, however, one or two points peculiarly noticeable about this barrow—first, the enormous quantity of burnt earth, suggesting that the funeral pyre was actually lit on the spot, which certainly was not the case in most of the other barrows, where the charcoal is only sprinkled here and there, or appears in the form of a small circular patch on the floor. Secondly, the two bands of charcoal, so full of osseous matter, would certainly go far to prove, what has been surmised by Bateman and others, that the slaves or prisoners were immolated at the decease of their master or conqueror.

Again, too, the different sizes and positions of the urns may, perhaps, indicate either degrees of relationship or rank of the persons buried. And this theory is somewhat corroborated by the contents. The central urn was examined on the spot, and, like all the others, with the exception of a round stone slightly indented, contained burnt earth, limy matter, and at the bottom the larger bones, which were less calcined, but which, owing to the want of proper means, we could not preserve. The other two were opened at the British Museum. At the bottom of the north-easternmost were also placed bones in a similar condition, amongst which Professor Owen recognized the *femur* and *radius* of an adult. The smallest urn also showed bones placed in the same manner at the bottom, but in this case smaller, and amongst them Professor Owen determined *processus dentatus*, and the body of the third cervical vertebra, and was of opinion that they were those of a person of small stature, or, perhaps, of a female. This is what might have been expected. And the fact of their being put in the smallest vessel, which, as we have noticed, was placed below the level of the others, certainly indicates a distinction made in the mode of burial of persons of either different ages or sexes.

The fact, too, that all the larger bones were placed by themselves at the bottom is worth noticing, and shows that they must have been carefully collected and separated from the burnt earth and charcoal of the pyre.

About another quarter of a mile off rise two more barrows, measuring exactly the same in circumference as the last, though not nearly so high, being raised only sixteen inches above the ground. Upon opening the southernmost, we soon came, on the east side, upon traces of charcoal, which increased to a bed of an inch and a half in thickness as we reached the centre. Here we found an urn of coarse pottery exactly similar in texture to those in the previous barrow. It was, however, in such a bad state of preservation, and so soft, from the wetness of the ground, that the furze-roots had grown through the sides, and it crumbled to bits on being touched. Some few pieces, however, near the bottom, we were able to preserve. Its shape, however, was well shown by the form which its contents had taken. It seems to have been, though much smaller, exactly of the same rude, straight-sided, and wide-mouthed pattern as the other urns, measuring seven inches in height, and in circumference, near the top, two feet two inches, and at the bottom, one foot four inches. The cast was composed entirely of burnt stones, and black earth, and osseous matter, reduced to lime, in which the furze-roots had imbedded themselves.

The fellow barrow, which was only about fifty yards distant, and whose measurements were exactly the same, contained also charcoal, though not in such large quantities, and fragments of an urn placed not in the centre, but near the extreme western edge. The remains here were in a still worse state of decomposition, and we could obtain no measurements, but only one or two pieces of ware, which, in their general coarseness and grittiness of texture, corresponded with the others, and not only showed their Keltic manufacture, but their extreme early date.[7]

This last mound, I may add, was composed of gravel, whilst the other was made simply of mould: and two depressions on the heath showed where the material had been obtained.

About two miles to the north-east, close to Ocknell Pond, lies a single barrow of much the same size as these two, though a great deal higher, being raised in the centre to three feet and a half. We began the excavation on the east side, proceeding to the centre, but found nothing except some charcoal, and peculiarly-shaped rolled flints, placed on the level of the ground.

We then made another trench from the north side, and close to some charcoal, about a foot and a half below the raised surface, came upon the neck of a Roman wine vessel (*ampulla*). Although we opened the whole of the east side, we could not find the remaining portion. The barrow bore no traces of having been previously explored, nor did the soil appear to have been moved. The fracture was certainly not recent, and it is very possible that some disappointed treasure-seekers in the Middle Ages had forestalled us, and time had obliterated all their marks in opening the mound.

From the position of the vessel at the top of the barrow, there had evidently been a second interment. The remains, however, are in accordance with what we might have expected. The barrow is situated not far from the Romano-Britishpotteries of Sloden, and close to it run great banks, known as the Row-ditch, marking, in all probability, the settlements of a Romano-British population.[8]

Neck of Roman Wine Vessel, Keltic Urn, and Flint Knives.

On Fritham Plain, not far from Gorely Bushes, lies another vast graveyard. The grave-circles are very similar in size to those round the large barrow on Bratley Plain, though a good deal higher, with, here and there, some oval mounds ranged side by side, as in a modern churchyard. In the autumn of 1862, I opened five of these, with the same result of finding charcoal in all, though placed in different parts, but in all instances resting on the natural ground, and giving evidence of only one interment. As in other cases, the grave-heaps were often alternately composed of mould and gravel. No traces of urns or celts were found, but in one or two a quantity of small circular stones, with indistinct marks of borings, which could hardly have accidentally collected.

About a quarter of a mile off, on the road to Whiteshoot,[9] lies, however, a square mound, measuring nine yards each way, and averaging a foot and a half in height. On opening it on the north side, we came upon the fragments of an urn, so much decayed, however, that we could only tell that they were, probably, Keltic.

On the west side, another trench, which had been made, showed the presence of charcoal, which kept increasing till we reached the centre, where we found what appeared to be the remains of three separate urns, placed in a triangle at about a yard apart. These also were in the same decayed state, and crumbled to pieces as we endeavoured to separate them from the soil. With some difficulty we managed to preserve a few fragments which were identical with those which had been previously discovered in the other barrows at Bratley. They contained, like most of the other vessels, burnt stones and white osseous matter reduced to lime. There seems, however, to have been some difference in their texture with that of the fragments found on the north side, which were less gritty and coarse, and which bore no traces of charcoal or lime.[10]

We will now leave Fritham, and cross Sloden and Amberwood Plantation. Not far from Amberwood Corner, and above Pitt's Enclosure, stand two barrows. The largest was opened thirty years ago by a labouring man, who, to use his own language, "constantly dreamt that he should there find a crock of gold." His opening was rewarded by discovering only some charcoal. In 1851, the Rev. J. Pemberton Bartlett also explored it with still less success. It is, however, a remarkable barrow, and differs in character from any of the preceding, being composed in the interior of large sub-angular flints, and cased on the outside with a rampart of earth. Beyond it lies another, very different in style, being made only of earth. This was also opened by Mr. Bartlett, who found some pieces of charcoal, and small fragments of a very coarsely-made urn.

About a mile away on Butt's Plain rise five more barrows, and beyond them again two more. Of the first five, two were explored by Mr. Bartlett, who was unsuccessful, and two by myself.

The two which I opened lie on the right of the track leading from Amberwood to the Fordingbridge road. The northernmost was considerably the largest, having a circumference of fifty yards, and was composed simply of gravel and earth. In it we found only a circle of charcoal placed nearly in the centre on the level of the ground.

The other was more remarkable. It measured only thirty yards in circumference, but was composed in the centre of raised earth, above which were piled large rolled flints, making a stratum of from two to three feet in depth on the sides, but gradually becoming thinner as it reached the centre, which was barely covered. It thus totally differed from that near Amberwood, where the earth flanked the stones instead of being the nucleus round which they were placed. In it we found a circle of charcoal ingrained with limy matter, a few remains of much calcined bones, and a fine stone hammer bored with two holes slantwise, to give a greater purchase to the handle.

Besides these, I opened a solitary barrow situated between Handycross Pond and Pinnock Wood, close to Akercombe Bottom. It measured twenty-seven yards in circumference, and three feet in height. After digging into it near the centre, we found in the white sand, of which the mound was chiefly composed, a good deal of charcoal on and below the level of the ground, but failed to discover any traces of an urn, although we went down to a considerable depth.

Further, a solitary oval mound stood on the south side of South Bentley, half way between it and Anses Wood. It measured two feet and a half in height, twelve yards in length, and seven in breadth. This also I opened, but failed to find even any remains of charcoal, and, from the easy-moving nature of the soil, am inclined to suspect that it was modern, and raised for some other purpose than that of burial. On the east side was a depression filled with water, from whence the soil was taken.

The most remarkable barrow, if it can be so called, in this part of the Forest, is at Black Bar, at the extreme west end of Linwood, measuring nearly four hundred yards in circumference, and rising to the height of forty feet or more. It is evidently in part factitious, for upon sinking a pit ten feet deep we reached charcoal mixed with Roman pottery, but not of a sepulchral character.

In its general appearance the mound is not unlike the famous Barney Barn's Hill, in Dibden Bottom, and close to it rises another, known as the Fir Pound, not much inferior in size. I made other openings on the top and sides, but discovered nothing further. To excavate it thoroughly would require an enormous time, and would in all probability not repay the labour. It looks, however, by the depressions on the summit, as if it had once been the site of Keltic dwellings. And this is in some measure corroborated by a small mound close to it, where, as if apparently left or thrown away, we found placed in a hole a small quantity of extremely coarse pottery—the coarsest and thickest which I have ever seen. Again, too, in a field close by, known as Blackheath Meadow, we everywhere met traces of Romano-British ware, very similar in shape and texture to that in Sloden, described in the next chapter.

The whole district just round here is most interesting. About a mile to the north is Latchmoor Stream and Latchmoor Green, marking, doubtless, some burial-ground; and not far off stands one of those elevated places, common in the Forest, with the misleading title of Castle.

I must not, too, forget to mention some barrows on Langley Heath, just outside the present eastern boundary of the Forest, and especially interesting from being situated so near to Calshot, where, as we have seen, Cerdic probably landed. Seven of them were opened by the Rev. J. Pemberton Bartlett. The mounds, averaging about twenty yards in circumference, were, in some cases, slightly raised, as much as a foot and a half, though in others nearly on a level with the natural surface of the soil. In them all was found a single grave, though, in one instance, two, running about three feet in depth, and containing only burnt earth and charcoal. They thus exactly corresponded, with the exception of the slight mound, with those on Bratley Plain.

With this we must conclude.[11] It would not be difficult to frame some theory from these results. I, however, here prefer to allow the simple facts to remain. As we have seen, the barrows in this part of the Forest, like all others of the same period, contained nothing, with the exception of the single stone-hammer, and the slinging pebbles, and the flake of flint, but nearly plain urns, full of only burnt earth, charcoal, and human bones. No iron, bronze, nor bone-work of any sort, was found, which would still further go to prove their extreme early age. Curiously enough, too, no teeth, bones, nor horn-cores of animals were discovered, as so often are in Keltic barrows.[12] Like all others, too, of an early date, there seem to have been several burials in the same grave, though this, as on Fritham Plain, is very far from being always the case. Some little regularity evidently prevailed with the different septs. Some, as at Bratley, placed the charred remains in a grave from two to three feet in depth; others, as at Butt's Plain, on the mere ground. On the other hand, a good deal of caprice seems to have been exercised as to the materials with which each barrow was formed, and the way and the shape in which it was built, as also the arrangement of the charcoal.

Further, perhaps, the different grades of life and relationship were marked by the presence and position of the urns. Whether this be so or no, it is certain that the mounds here which contained mortuary vessels were, as a rule, more elevated, and in nearly all instances placed by themselves. The fact, too, of the cube-shaped mound with its remains of four urns should be kept in mind.

Little more can with certainty be said. The flint knives which have been picked up in the Forest, the stone hammer in the grave, the clumsy form and make of the urns, the places, too, of burial—in the wide furzy Ytene, in after-times the Bratleys, and

Burleys, and Oakleys, of the West-Saxons—all show a people whose living was gained rather by hunting than agriculture or commerce.

Barrows on Beaulieu Plain.

Footnotes

1. I may as well add that a little way from where the Bound Oak formerly stood, near Dibden, and between it and Sandy Hill, lies a small mound, thirty yards in circumference, and three feet high in the centre, surrounded by an irregular moat, from which the earth had been taken. This I opened in 1862, driving a broad trench from the east to the centre, and another from the south to the centre, which, as also the west side, we entirely excavated; digging below the natural soil to the depth of four feet. Nothing, however, was found, though I have no doubt charcoal was somewhere present.
Beyond this, in Dibden Bottom, rises a large mound, from twenty to thirty feet high, apparently of a sepulchral character, known as Barney Barns Hill. Proceeding, close to Butt's Ash End Lane, and near the Roman, or rather British, road to Leap (see chap, v.,

p. 56), stand two barrows, the northernmost one hundred and the southernmost eighty yards in circumference. Farther away, in Holbury Purlieu, are three more, each with a circle of about seventy yards. To the west of these, in the Forest, as shown in the illustration at page 213, rise four more, the three farthest forming a triangle. Beyond these, again, about three-quarters of a mile distant, near Stoneyford Pond, lie four others, respectively ninety, one hundred, and seventy yards in circumference. To the north rise three more, known as the Nodes; the westernmost about one hundred yards in circumference; the other two, which are ovaler and form twin barrows, being one hundred and fifty and one hundred yards. Two more stand on the side of the Beaulieu road to Fawley. All these, with others on Lymington Common and near Ashurst Lodge, and on the East Fritham Plain, still remain to be explored. For the barrows opened by the Rev. J. Pemberton Bartlett, on Langley Heath, see farther on, page 211.
2. *South Western Parts of Hampshire*, vol. i. pp. 69-79.
3. Warner probably meant an overhanging brim, such as is common to most of the early Keltic cinerary urns, or, perhaps, one like that of the left-hand urn in the illustration at p. 196, which is more contracted than the others. He unfortunately gives us no dimensions.
4. This camp was probably, since coins of Claudius have been found, occupied by Vespasian, when he conquered the Isle of Wight. A bronze celt was found here some eighty years ago, and came into the possession of Warner. Others have been discovered, in great quantities, in various parts of the Forest, two of which are engraved in *Archæologia*, vol. v., plate viii., figs. 9 and 10. Brander, too, the well-known antiquary, found others at Hinton, on the west border of the Forest (*Archæologia*, vol. v. p. 115). Mr. Drayson has also picked up two flint knives at Eyeworth, which are figured, showing both the under and upper surfaces, at p. 206.
5. As in Derbyshire all barrows are marked by the terminal

low—*hlæw*, a grave, so in the Forest they seem particularized by a reference to the Old-English *lie*. Thus, near the Beaulieu barrows we find Lytton Copse and Common, and at the west end of the Forest, not far from Amberwood, meet another Latchmoor. I may notice that just outside the Forest, in Darrat's Lane—a word which often occurs—we find a place, near some mounds, called "Brands," equivalent to the "Brund" of Derbyshire, and having reference to the burning funeral pyre. (See Bateman's *Ten Years' Diggings*, Appendix, p. 290.)

6. I certainly think that these urns were fired, though imperfectly. As Mr. Bateman remarks, sun-baked specimens soon return to their original clay. See Appendix to *Ten Years' Diggings*, p. 280. These three urns, with all the other fragments of cinerary vessels found in the Forest, I have placed in the British Museum, where they have been restored. The artist has represented them exactly as they appeared on the second day of digging. The fractures in the central urn were caused by an unlucky blow from a pick-axe. The measurements are as follows:—

The north-eastern urn—	Circumference at	top	3 ft.
„	„	bottom	1 „ 6 in.
„	Total height		1 „ 4½ „
The central urn—The same.			
The south-western urn—	Circumference at	top	2 „ 9 „
„	„	bottom	1 „ 4½ „

„	Total height		1 „ 1¾ „

7. I am inclined to think that here, as in the similar instance on Fritham Plain, the urns were put in the mound entire, and not, as is sometimes the case, in fragments. The pieces had no appearance of being burnt after the fractures had taken place, which were here simply the result of decay. See on this point Bateman's *Ten Years' Diggings*, pp. 191, 192, where Mr. Keller's letter to Sir Henry Ellis on the subject is given.
8. Instances have been known where the top of a Roman cinerary urn has been taken off, and replaced; but, from the narrowness of the neck, I hardly think this vessel was used for such a purpose. I give with it also a late British urn found, some twenty years ago, in a barrow outside the present Forest boundary, in a field known as Hilly Accombs, near Darrat's Lane, which has been previously mentioned. It measures 6 inches in height, and has a circumference of 1 foot 9 inches round the top, and 1 foot at the base. With it was discovered another, but I have been unable to learn in whose possession it now is, or what has become of the Roman glass unguent bottle found in Denney Walk (see the *Antiquities of the Priory of Christchurch*, by B. Ferrey and E. W. Brayley, p. 2, foot-note). The two flint knives were discovered by Mr. Drayson, near Eyeworth Wood, and somewhat resemble the chipping found in the largest barrow at Bratley, and were, perhaps, cotemporary. The celts found by Warner and Brander, with others in the possession of Gough, mentioned at p. 199, foot-note, were bronze. Mr. Keeping also discovered a stone celt in the drift, but this belonged to a far earlier period. See chap. xix.
9. There are two large heathy tracts known as Fritham Plain; the one to the east, where stand several large trenched barrows, which still remain to be opened; and the West Plain, where these excavations took place.

10. An attempt to examine this barrow had been previously made, but the explorers had opened a little to the south-west of the spot where the pottery lay. It is just possible that the large square in Sloden may be of the same character. I cut a small opening at the western end, but it is impossible, on account of the trees, to make any satisfactory excavation. Whatever might have been its original purpose, it was certainly never the site of a church, as is commonly supposed. Sec ch. iii., p, 32, foot-note.
11. To assist the archaeologist, I have marked on the map the sites of all the barrows of which I am aware. In the British Museum is a small urn, found in a barrow at Broughton, on the borders of Hampshire, about twelve miles north of the Forest, measuring three inches in height, and, though so much less, somewhat resembling, with its two small ears, as also in the general character and texture of its ware, those found in the Bratley barrow. The Rev. J. Compton also informs me that some years ago a plain urn was discovered in a barrow on his father's property at Minestead, in the Forest. I hear, too, that other urns have been found in barrows near Burley on the west, and near Butt's Ash Lane on the east side of the Forest, but they have long ago been lost or destroyed, and I am unable to learn even their general form. I trust, therefore, permission will not be granted to open the mounds which are unexplored, except to those who can produce some credentials that they are fitted for the task, and are doing it from no idle curiosity, but legitimate motives. Too much harm has been already done, and too many barrows have been already rifled, without any record being made of their contents. Nearly all that we know of Kelt or Old-English we learn from their deaths. Their history is buried in their graves.
12. In Mr. Birch's *Ancient Pottery*, vol. ii. pp. 382, 383, will be found a list of the notices of the various discoveries of Keltic urns, scattered through the different Archaeological Journals and Collections, which will save the student much time and labour. A

most valuable paper on the subject, by Kemble, was published in the *Archæological Journal*, vol. xii. number 48, p. 309.

CHAPTER XVIII.

THE ROMAN AND ROMANO-BRITISH POTTERIES.

Wine-Flask, Drinking-Cups, and Bowls.

FROM time to time the labourer, in draining or planting in the Forest, digs down upon pieces of earthenware, whilst in the turfy spots the mole throws up the black fragments in her mound of earth. The names, too, of Crockle—Crock Kiln—and Panshard Hill, have from time immemorial marked the site of at least two potteries. Yet even these had escaped all notice until Mr. Bartlett, in 1853, gave an account of his excavations, and showed the large scale on which the Romans carried on their works, and the beauty of their commonest forms and shapes.[1]

Since then both Mr. Bartlett and myself have at different times opened various other sites, and some short notice of their contents may, perhaps, not be without interest.

Fifty years ago, when digging the holes for the gate-posts at the south-west corner of Anderwood Enclosure, the workmen discovered some perfect urns and vases. These have, of course, long since been lost. But as the place was so far distant from the potteries at Crockle, I determined to re-open it. The site, however, had been much disturbed. Enough though could be seen to show that there had once been a small kiln, round which were scattered for three or four yards, in a black mould of about a foot and a half in depth, the rims, and handles, and bottoms of vessels of Romano-British ware. The specimens were entirely confined to the commonest forms, all ornamentation being absent, and the ware itself of a very coarse kind, the paste being grey and gritty.

About a mile and a half off, in Oakley Enclosure, close to the Bound Beech, I was, however, more fortunate. Here the kiln was perfect. It was circular, and measured six yards in circumference, its shape being well-defined by small hand-formed masses of red brick-earth. The floor, about two feet below the natural surface of the ground, was paved with a layer of sand-stones, some of them cut into a circular shape, so as to fit the kiln, the upper surfaces being tooled, whilst the under remained in their original state. As at Anderwood, the ware was broken into small fragments, and was scattered round the kiln for five or six yards. The specimens were here, too, of the coarsest kind, principally pieces of bowls and shallow dishes, and, perhaps, though of a different age, not so unlike as might at first sight be supposed to the

"Sympuvium Numæ, nigrumque catinum,
Et Vaticano fragiles de monte patellæ."

These appear to be the only kilns which, perhaps from the unfitness of the clay, were worked in this part of the Forest, and were used only in manufacturing the most necessary utensils in daily life.

Of far greater extent are the works at Sloden, covering several acres. All that remains of these, too, are, I am sorry to say, mere fragments of a coarse black earthenware. And although I opened the ground at various points, I never could meet with anything perfect. Yet the spot is not without great interest. The character and nature of the south-western slope exactly coincide with Colt Hoare's description of Knook Down and the Stockton Works.[2] Here are the same irregularities in the ground, the same black mould, the same coarse pottery, the same banks, and mounds, and entrenchments, all indicating the settlement of a Romano-British population. Half-way down the hill, not far from two large mounds marking the sites of kilns, stretch trenches and banks showing the spaces within which, perhaps, the potters' huts stood, or where the cultivated fields lay, whilst at one place five banks meet in a point, and between two of them appear some slight traces of what may have been a road.[3]

At the bottom of the hill, but more to the south-westward, stands the Lower Hat, where the same coarse ware covers the earth, and where the presence of nettles and chickweed shows that the place has once been inhabited.

The Crockle and Island Thorn potteries lie about a mile to the north-east. At Crockle there were, before Mr. Bartlett opened them, three mounds, varying in circumference from one hundred and eighty to seventy yards, each, as I have ascertained, containing at least three or four, but probably more, kilns. As the lowest part of the smallest and easternmost mound had not been entirely explored, I determined to open this piece. Beginning at the extremity, we soon came upon a kiln, which, like the others discovered by Mr. Bartlett, only showed its presence by the crumbling red brick earth. An enormous old oak-stump had grown close beside it, and around the bole were heaped the drinking-vessels and oil-flasks, which its now rotten roots had once pierced.

Necks of Oil-Flasks.

Nothing could better show, as the excavation proceeded, the former state of the works. Here were imbedded in the stiff yellow putty-like clay, of which they were made, masses of earthenware, the charcoal, with which they were fired, still sticking to their sides—pieces of vitreous-looking slag, and a grey line of cinders mixed with the red brick earth of the kiln. The ware remained just as it was cast aside by the potter. You might tell by the bulging of the sides, and the bright metallic glaze of the vessels, how the workman had overheated the kiln;—see, too, by the crookedness of the lines, where his hand had missed its stroke. All was here. The potter's finger-marks were still stamped upon the bricks. Here lay the brass coin which he had dropped, and the tool he had forgotten, and the plank upon which he had tempered the clay.[4]

Necks of Wine-Vessels and Oil-Flask.

The Island Thorn potteries had been so thoroughly opened by Mr. Bartlett, that I there made but little further explorations, and must refer my readers to his account,[5] only here adding that the ware scarcely differed, except in shape and patterns, from that at Crockle.

About a mile westward stands Pitt's Enclosure, where in three different places rise low mounds, two of which, since the publication of his account, have been opened by Mr. Bartlett, but from which he only obtained fragments.

The third, which I explored in 1862, was remarkable for the number of kilns placed close together, separated from each other by only mounds of the natural soil. In all, there were five, ranged in a semicircle, and paved with irregular masses of sandstone. They appear to have been used at the time at which they were left for firing different sorts of ware. Close to the westernmost kiln, we found only the necks of various unguent bottles, whilst the easternmost oven seems to have been employed in baking only a coarse red panchion, on which a cover (*operculum*), with a slight knob for a handle, fitted. Of these last we discovered an enormous quantity, apparently flung away into a deep hole.

Near the central kilns we found one or two new shapes and patterns, but they were, I am sorry to say, very much broken, the ware not being equal in strength or fineness to that at Crockle. The most interesting discovery, however, were two distinct heaps of white and fawn-coloured clay and red earth, placed ready for mixing, and a third of the two worked together, fit for the immediate use of the potter.

Near to these works stretch, on a smaller scale, the same embankments which mark the Sloden potteries. One is particularly noticeable, measuring twenty-two feet in width, and running in the shape of the letter Z. In the central portion I cut two trenches, but could discover nothing but a circle of charcoal, looking as if it was the remains of a workman's fire, placed on the level of the natural soil. Another trench I opened at the extreme end, as also various pits near the embankment, hut failed to find anything further.

At Ashley Rails, also, close by, stand two more mounds, which cover the remains of more ware. These I only very partially opened, for the black mould was very shallow, and the specimens the same which I had found in Pitt's Wood.

Besides these, there are, as mentioned in the last chapter, extensive works at Black Heath Meadow at the west-end of Linwood, but they are entirely, like those in Sloden, Oakley, and Anderwood, confined to the manufacture of coarse Romano-British pottery. This last ware seems to differ very little in character or form. The same shapes of jars (copied from the Roman *lagenæ*) were found by Mr. Kell near Barnes Chine in the Isle of Wight,[6] though at Black Heath, as in the other places in the Forest, handles, through which cords were probably intended to pass, with flat dishes, and saucer-like vessels (shaped similar to *pateræ*), all, however, in fragments, occurred.[7]

Such is a brief account of the potteries in the Forest. Their extent was, with two exceptions, restricted to one district, where the Lower Bagshot Sands, with their clays, crop out, and to the very same bed which the potters at Alderholt, on the other side of the Avon, still at this hour work.

The two exceptions at Oakley and Anderwood are situated just at the junction of the Upper Bagshot Sands and the Barton

Clays, which did not suit so well, and where the potteries are very much smaller, and the ware coarser and grittier.

The date of the Crockle potteries may be roughly guessed by the coins, found there by Mr. Bartlett, of Victorinus.[8] These were much worn, and, as Mr. Akerman suggests, might be lost about the end of the third century; but the potteries were probably worked till or even after the Romans abandoned the island.

There is nothing to indicate any sudden removal, but, on the contrary, everything shows that the works were by degrees stopped, and the population gradually withdrew. None of the vessels are quite perfect, but are what are technically known as "wasters." The most complete have some slight flaw, and are evidently the refuse, which the potter did not think fit for the market.

The size of the works need excite no surprise, when we remember how much earthenware was used in daily life by the Romans—for their floors, and drinking-cups, and oil and wine flasks, and unguent vessels, and cinerary urns, and boxes for money. The beauty, however, of the forms, even if it does not approach that of the Upchurch and Castor pottery, should be noticed. The flowing lines, the scroll-work patterns, the narrow necks of the wine-flasks and unguent vessels, all show how well the true artist understands that it is the real perfection of Art to make beauty ever the handmaid of use.

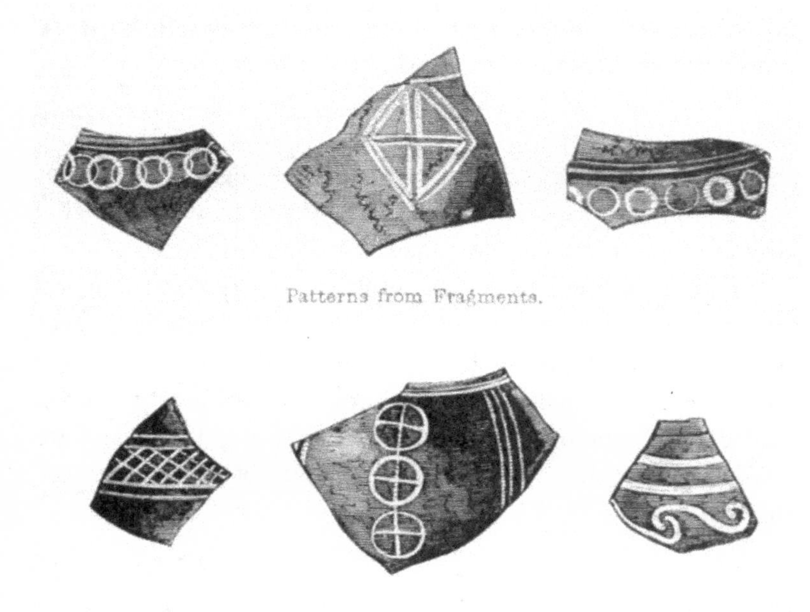

Patterns from Fragments.

Patterns from Fragments.

Another thing, too, is worthy of notice, that the artist was evidently unfettered by any given pattern or rule. Whatever device or form was at the moment uppermost in his mind, that he carried out, his hand following the bent of his fancy. Hence the endless variety of patterns and forms. No two vessels are exactly alike. In modern manufactures, however, the smooth uniformity of ugliness most admirably keeps down any symptoms of the prodigal luxuriance of beauty.[9]

We must, however, carefully beware of founding any theory, from the existence of these potteries, that the Forest must therefore have been cultivated in the days of the Conqueror. The reason why the Romans chose the Forest is obvious,—not from its fertility, but because it supplied the wood to fire the kilns; the same cause which, centuries after, made Yarranton select Ringwood for his smelting-furnaces. We must, too, bear in mind that after the Romans abandoned the island the natives soon went back to their primitive state of semi-barbarism; and further, that the interval between the Roman occupation and the Norman Conquest was nearly as great as that between ourselves and the Conqueror—a period long enough for the Kelts, and West-Saxons, and Danes to have swept away in their feuds all traces of civilization.

But what we should see in them is that beauty of form, which in simple outline has seldom been excelled, proclaiming a people who should in their descendants be the future masters of Art, as then they were of warfare.

The history of a nation may be plainer read by its manufactures than by its laws or constitution. Its true aesthetic life, too, should be determined not so much by its list of poets or painters, as by the beauty of the articles in daily use.

And so still at Alderholt, not many miles off, the same beds of clay are worked, and jars, and flasks, and dishes made, but with a difference which may, perhaps, enable us to understand our inferiority in Art to the former rulers of our island.

What further we should see in the whole district, is the way in which the Romans stamped their iron rule upon every land which they conquered. Everywhere in the Forest remain their traces. Urns, made at these potteries, full of their coins, have been dug up at Anderwood and Canterton. Iron nails at Cadenham, millstones at Studley Head, bricks at Bentley, iron slag at Sloden, with the long range of embankments stretching from wood to wood, and the camps at Buckland Rings and Eyeworth, show that they well knew both how to conquer in war and to rule in peace.

Oil-Flask. Drinking-Cups, Bowl, and Jar.

Footnotes

1. *Archæologia*, vol. xxxv. pp. 91-96.
2. See, too, Mr. Carrington's "Account of a Romano-British Settlement near Wetton, Staffordshire," in Bateman's *Ten Years' Diggings*, pp. 194-200. I have never found any stone floors, but this may be accounted for by the difficulty of procuring paving-stones in the district. The best guide which I know for discovering any ancient settlements is the presence of nettles and chickweed, which, like the American "Jersey-weed," always accompany the footsteps of man. These plants are very conspicuous in the lower parts of Sloden, as also at the Crockle and Island Thorn potteries.
3. The spot where these banks intersect each other is known as Sloden Hole, and is well worthy of notice. The annexed plan will best show the character of the place.

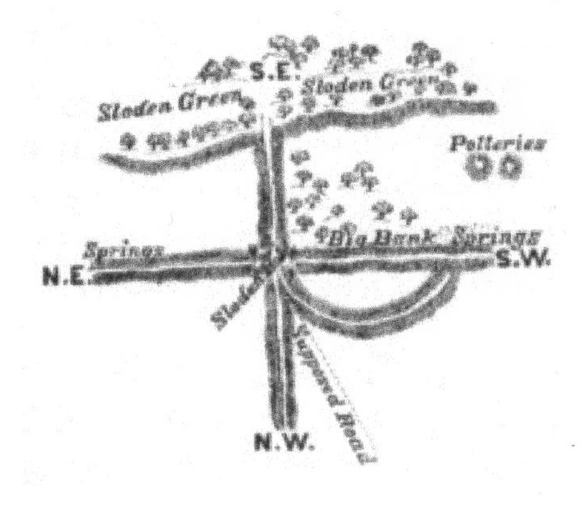

The largest bank is that which runs to the south-west, measuring four yards across, and proving by its massiveness that it is a Roman work. Upon digging, as shown in the plan, at the point of intersection, we found pieces of iron and iron slag, sandstone, charcoal, and Roman pottery similar to that made in Crockle. Many of these banks run for long distances. That to the south-east reaches the top of Sloden Green, about half a mile off, whilst the north-east bank stretches for nearly a mile to Whiteshoot. There are, too, other banks scattered about Sloden, which, if examined, would doubtless yield similar results, but none are so well defined as these. The largest bank which I know in the district stretches from Pitt's Enclosure, in a south-easterly direction across Anderwood, and so through the southern parts of Sloden.

4. The most noticeable specimens which I discovered were a strainer or colander, a funnel, some fragments of "mock Samian" ware; part of a lamp, with the holes to admit air, as also for suspension; and some beads of Kimmeridge clay, proving, by being found here, their Roman origin. The iron tools of the workmen had been dropped into the furnace, and were a good deal melted. The wood owed its preservation to the ferruginous soil in which it was imbedded, and was in a semi-fossilized state.

Nothing less slight than a plank could have lasted so long. The fingermarks and portion of the hand were very plain on one of the masses of brick-earth. The coin, I am sorry to say, is too much worn to be recognized. These, with the other vessels, *pateræ*, *urceoli*, *lagenæ*, *pocula*, *acetabula*, &c, I have placed in the British Museum, where is also Mr. Bartlett's rich collection. The patterns, with the necks of *ampullæ* and *gutti*, as also the specimens at pages 214, 225, will, I trust, give some general idea of the beauty of the ware, and can be compared with those given by Mr. Akerman in *Archæologia*, vol. xxxv. p. 96, and by Mr. Franks in the *Archæological Journal*, vol. x. p. 8. The commonest shape for a drinking-vessel is the right-hand figure at page 225, known in the Forest, from the depressions made by the workman's thumb, as a "thumb pot." It is sometimes met with considerably ornamented, and varies in height from ten to three inches. The principal part of the pottery is slate-coloured and grey, and faint yellow, but some of a fine red bronze and morone, caused by the overheating of the ovens. The patterns are thrown up by some white pigment, though a great many are left untouched by anything but the workman's tool. When chipped, the ware, by being so well burnt, is quite siliceous. The so-called crockery of the southern part of the Forest is nothing else but the plates of turtles imbedded in the Freshwater marls. I find I was misinformed with regard to the recent discovery of a Roman glass manufactory at Buckholt, mentioned in chapter v., page 51, footnote. Some most interesting glass-works, however, the earliest known in England, dating from the fourteenth century, occur at Buckholt in Wiltshire, nine miles from Salisbury, and were explored by the Rev. E. Kell, F.S.A. See *Journal of the Archæological Association*, 1861, vol. xvii pp. 55-70.

5. *Archæologia*, vol. xxxv. pp. 95, 96.

6. See *Journal of the Archæological Association*, vol. xii. pp. 141-145, where some figures of the jars are given.

7. In Eyeworth Wood I have found pieces of Roman wine and oil

flasks, but they were left here by the former inhabitants, and not made on the spot. The place known as Church Green is evidently the site of a habitation. In the autumn of 1862 I made several excavations; but there was some difficulty attending the work, as the ground had been previously explored by the late Mr. Lewis, the author of the *Historical Inquiries on the State of the New Forest*. The evidence, however, of the Roman pottery was sufficient to show its occupation during the Roman period, and to dispel the illusion that it was ever the site of a church. On the north-east side of the wood are the remains of a fine Roman camp, the *agger* and *vallum* being in one place nearly complete.

8. I may add that Mr. Drayson also possesses coins of Victorinus, and Claudius Gothicus, found in various parts of the Forest, the last in one of the "thumb-pots," with 1700 others, perhaps, indicating the period when the Crockle and Island Thorn Potteries were in their most flourishing condition.

9. In *Archæologia*, vol. xxxv. p. 99, Mr. Akerman has given a series of patterns, which show the variety of designs used according to the fancy of each workman. The pattern on the right-hand side of our second illustration at p. 223 is used as a border in the toga of the later Roman empire. The height of the wine vessel at p. 214 is seven inches and a half; of the oil-flask at p. 225, five inches; of the largest drinking cup, five inches; and the smallest, three inches and three-quarters; the jar, two inches.

CHAPTER XIX.

PARISH REGISTERS AND CHURCHWARDENS' BOOKS.

Boldre Church

As the monasteries of former days preserved the general records of the times, so, in a minor degree, do our churches preserve the special history of our villages. In the social life of the past our Church Books are the counterpart of our Corporation Books, performing quite as much for their own parishes as the latter for their boroughs; not only giving, in the register, a yearly census of the population, but by the Churchwardens' Accounts the social and religious life of each period.

Added to this also the clergyman, having nowhere else to chronicle them, has often entered in his register the passing events of the day; so that this further possesses, at times, a wider historical interest than could have been expected, giving us often glimpses of the views of men, who, however unsympathetic with the changes and fortunes of the hour, still carry, from their office and position, some not inconsiderable weight.

All these books are far too seldom consulted. The few notes we shall make are by no means given as examples of what may be elsewhere found, but must be looked upon only as extracts from the books of a district, where we naturally could expect little of any general interest.

The New Forest has never been, since registers became the law of the land, the scene of any of the great events of English history—never the theatre of the Civil Wars, as the Midland Counties, where entries of victories and defeats, and battles and sieges, are mixed with the burials and births.

Various causes, too, especially the scanty and scattered population, have contributed to the late date at which nearly all the Forest registers commence.[1] Still, at Eling, there occurs the second earliest parish register in Hampshire, beginning one year before Cromwell's Act has been passed; showing, as was before noticed, that this part of the Forest was always the richest, and, consequently, the most civilized.[2] In this register we find the following most interesting entry:—

"1654. Thomas Burges, the sonne of William Burges and Elizabeth Russel, the daughter of Elizabeth, the now wife of Stephen Newland, were asked three Sabbath dayes, in the Parish Church of Eling: sc: Apriel 16th, Ap^{r} 23rd, Ap^{r} 30th, and were marr: by Richard Ld Cromwell, May xxii^{d}."

I need scarcely add that it was under the Protector that an Act of Parliament was passed in 1653, enabling any persons, after the due proclamation of the banns in the church or chapel, or in the market-place, on three market days, to be married by a simple affirmation before a magistrate; thus in a remarkable way nearly anticipating modern legislature.[3] The Protector's son, at the date of this entry, was probably living at Hursley, about ten miles away to the north.

Going across to the other side of the Forest, we shall, at Ellingham, find, in the Churchwardens' Books, an entry in a different way quite as interesting. The leaf is, I am sorry to say, very much torn, and, towards the lower part, half of it is wanting. I give, however, the extract as it stands, indicating the missing passages by the breaks:—

"Martii 13. Anno dom. 1634. A special license, granted by the moste reverende ffather in God, William Lord Archbishop of Canterbury his Grace, under his Grace's hand and seale, used in the like grants, dated the nyneteenth day of ffebruarie, Anno dom. 1634, and second yeare of his Grace's translation. And confirmed by the Letters patents of our Sovraigne Lord Charles the King's ma.tie that now is .... Under the Greate Seale of England ffor Sr White Beconsaw of this parish and county of Southton .... (and) Dame Edith hys wife ffor the tyme of their naturell (lives) .... to eate flesh on the daies p'hibited by the Lawe .... (upon condition of their giving to the) poore of the p'ish .... Thirteene shillings ...."

Whether or no the knyght and his lady were to give the sum yearly, as seems most probable, it is impossible, from the torn condition of the leaf, to say. Their daughter was the noble Alice Lisle. The licence, of course, refers to the prohibition against eating meat on Fridays and Saturdays, and other specified times, first made by Elizabeth for the encouragement of the English fisheries, which had even in her reign begun to decay.[4] And now that we are on the subject of Churchwardens' Books, let me give some brief extracts from those of Ellingham:—

"1556. Itm for waxe ix*d.*
tm for a gyrdle iij*d.*
Itm for waxe and for makynge of ye paschall and fontetapers
xv*d.*
First payed for a rod (rood) xij*s.*
Itm payed for the paschall and fontetapers ij*s.* viij*d.*
"1558. First payed for the pascall and fontetapers xxij*d.*
Itm payed for frankeincense i*d.*"

Such notices well prove how quick and strong was the reaction from Protestantism to Catholicism when favoured by the State. Again, to still further show the variety of entries, let me make some extracts from the Fordingbridge Churchwardens' Books:—

"1636. I^tm^ for a fox-head 0 1^s^ 0
I^tm^ for one badgers head 0 1 0
I^tm^ for one fox-head 0 1 0"

Among miscellaneous notices, as giving the average wages of the day, and the prices of various articles, let me add also the following from the same accounts:—

"1609. I^{tm} laide out for a pint of muskadine vii^{d}"
"1616. I^{t} for viij dayes' worke for three men xxiij^{s}
I^{t} for a new beel-Rope iij^{s} iiij^{d}
It for a daye's worke for three men iij^{s} iij^{d}
I^{t} for a booke of artykeels iij^{s}
I^{t} for mates (mats) about the Communyon tabelle xiij^{d}

I^{t} payde the Person for keeping the Stocke iij^{s} iiij^{d}"

These accounts, too, like all others, are full of items for the repairs of the bells and bell-ropes, confirming what may be found in the narratives of old French and Italian travellers concerning our English passion for bell-ringing. The following looks very much like cause and effect:—

"1636. It~m to the Ringers one ye Kinges daye. ij^{s} vj^{d}
It~m for one belroape i^{s} iv^{d}"

The "King's day" was that on which the King ascended the throne. Again, to show the mixed and varied contents of the Churchwardens' Books, we will once more go back to those of Ellingham. Under the date of 1556 we find:—
"It~m for a baudericke of the great bell xij^{d}
It~m for a lanterne viij^{d}
It~m for nailes and sope iij^{d}"

Under the head of "Layinges out in the secunde yere," meaning

1557, we meet:—

"It~m for a pot of claye iij^{d}
It~m payed for ij bokes x^{s}
It~m payed for smoke sylver ij^{s} xi^{d}"

And, again, under the "Layinges out in the thyrdde yere," we find:—

"It~m payed for storynge of the tythynge harnesse xviij^{d}
It~m for white lether iij^{d}
It~m for lyme and vj creste tyles xxi^{d}
Itm for surplus for the clerke (clergyman) iij^{s}
Itm for smoke silvar xvij^{d}"

All these entries, to the church historian, and no less to the general student, cannot be without peculiar interest. The smoke silver, which so frequently occurs, is either the money paid for certain privileges of cutting fuel, which, as we have seen, was formerly the case in the Forest, or an assessment on the houses according to the number of hearths, but more probably the former.[5] The general reader will scarcely care for more, but I trust elsewhere to give further extracts from these most interesting books.

Turning back to the Registers, let me add from the Ibbesley Parish Register Book, as so few people have seen a specimen, an entry of an affidavit of burial in a woollen shroud, in compliance with the Act passed in 1679, for the encouragement of the woollen manufacture in England.[6] It thus runs, placed opposite to the entry of the person's burial, and written in the same handwriting:—"Jan. 9th, 1678/79, I rec[d] a certificate from Mr. Roger Clavell, Justice of y[e] peace at Brokenhurst, that Thomas King and Anthony King, sons of Anthony King, deceased, did make oath before him, the sayd Roger Clavell, that the aforesayd Antony King was buried according to the late Act of Parliament."

And again, opposite to the entries of their deaths, we find—"November 11th.—Certified by John Torbuck, Vicar of Ellingham, y[t] Edward Baily and Nicholas Baily, of Ibsely, were buried in woollen only."

Pope's lines on Mrs. Oldfield need hardly here be quoted. To conclude, of the parish books in the district let me only say that at Fordingbridge may be found an inventory of all the church furniture for 1554; at Christchurch, notes of a Papist buried by women, for no one else would place her in the grave; and entries of lay marriages; at Ibbesley, lists of collections "towards the redemption of the poor slaves out of Turkey," "for the poor French Protestants," "for the redemption of captives," and "for the distressed Protestants beyond the sea,"—all testifying to the social and moral condition of the people, without which it is impossible to give the history of any district or any country.

The Norman Font in Brockenhurst Church.

Footnotes

1. The following dates prior to 1700 of the Parish Registers in the Forest district are taken from the *Parish Register Abstract:* Accounts and Papers: 1833, vol. xxviii (No. 13), p. 398:—

Eling	1537	|	Milton	1654
Christchurch	1586	|	Lymington	1662
Milford	1594	|	Dibden	1665
Boldre	1596	|	Fawley	1673

<table>
<tr><td>Ellingham</td><td>1596</td><td>|</td><td>Breamore</td><td>1675</td></tr>
<tr><td>Bramshaw (loose leaves)</td><td>1598</td><td>|</td><td>Sopley</td><td>1678</td></tr>
<tr><td>Fordingbridge</td><td>1642</td><td>|</td><td>Minestead</td><td>1682</td></tr>
<tr><td>Beaulieu</td><td>1654</td><td>|</td><td>Ringwood</td><td>1692</td></tr>
<tr><td>Ibbesley</td><td>1654</td><td>|</td><td>Brockenhurst</td><td>1693</td></tr>
</table>

2 See chapter v., p. 51, *foot-note.*
3. Part of the Act is quoted in Burn's *History of Parish Registers*, second edition, pp. 26 and 27, and where, at pp. 159, 160, 161, are given several examples of this kind of marriage—amongst them, that of Oliver Cromwell's daughter Frances, in 1657, from the Register of St. Martin's-in-the Fields.
4. Burn, in his *History of Parish Registers*, second edition, pp. 171, 172, 173, gives several similar instances of such licences. These most valuable books at Ellingham are, notwithstanding the incumbent's care, in a shocking state of preservation. I trust some transcript of them may be made before they quite fall to pieces. Ellingham also possesses another book containing the names of the owners of the different pews in the church in 1672, invaluable to any local historian. In the beginning of this book are inserted a number of law-forms of agreements, wills, and indentures, probably for the use of the clergyman, who was, perhaps, consulted by his parishioners in worldly as also spiritual matters. In the Register there is, unfortunately, no mention of the death of Alice Lisle, as the burials are torn out from 1664 to 1695.
5. See *Notes and Queries*. First Series, vol. ii., pp. 344, 345. In the Churchwardens' Books of Fordingbridge we find—"1609. For

smokemony, for makynge and deliveringe of the bills xvjd," which would confirm the first explanation given in the text.

6. 30 Car. II., cap. iii. See *Journals of the House of Commons*, vol. viii., p. 650; ix., p. 440. In Burn's *History of Parish Registers*, second edition, p. 117, may be found a much more complicated affidavit than those given in the text.

CHAPTER XX.

THE GEOLOGY.

The Barton Cliffs

I have endeavoured, whenever there was an opportunity, to point out the natural history of the Forest, feeling sure that, from a lack of this knowledge, so many miss the real charms of the country. "One green field is like another green field," cried Johnson. Nothing can be so untrue. No two fields are ever the same. A brook flowing through the one, a narrow strip of chalk intersecting the other, will make them as different as Perthshire from Essex. Even Socrates could say in the *Phædrus*, τὰ μὲν οὖν χωρία καὶ τὰ δένδρα οὐδέν μ' ἐθέλει διδάσκειν· and this arose from the state, or rather absence, of all Natural Science at Athens. Had that been different he would have spoken otherwise.

The world is another place to the man who knows, and to the man who is ignorant of Natural History. To the one the earth is full of a thousand significations, to the other meaningless.

First of all, then, for a few words on the geology of the Forest; for upon this everything depends—not only the scenery, but its Flora and Fauna, the growth of its trees and the course of its streams. Throughout it is composed of the Middle-Eocene, the Osborne and Headon Beds capping the central portion, with their fluvio-marine formation. The Upper Bagshot develops itself below them, and is succeeded by the Barton Clays, so well exposed on the coast, and finally by the Bracklesham Beds, which crop out in the valley of Canterton, trending in a south-easterly direction to Dibden.

Here, then, where the New Forest stands, in the Eocene period, rolled an inland sea, whose waves lashed the Wiltshire chalk hills on the north, moulding, with every stroke of their breakers, its chalk flints into pebbles, dashing them against its cliffs, as the waves do at this very hour those very same pebbles along the Hurst beach. Its south-western boundary-line between Ballard Head and the Needles was rent asunder by volcanic action, and the chalk-flints flung up vertically mark to this day the violence of the disruption.

Long after this the Isle of Wight was altogether separated by the Solent from the mainland, but still ages before the historic period. The various traditions, as to the former depth of the channel, how Sir Bevis, of Southampton, waded across it, how, too, the carts brought the Binstead stone for building Beaulieu Abbey over the dry bed at low water, have been previously given. The passage, too, in Diodorus Siculus has been already examined,[1] and there can be no doubt, notwithstanding his also making it, like the traditions, a peninsula at low water, that his Ictis is the Isle of Wight and not St. Michael's Mount. The mere local evidence of the mass of tin, the British road—more like a deep trench than a road—still plainly traceable across the Forest, the names along it corresponding with that of its continuation in the Island, would alone, most assuredly, show that this was the place whence the first traders, and, in after-times, the Romans, exported their tin. We must, however, remember that the channel of the Solent was caused by depression rather than by excavation; and that at this moment an alteration in the levels, as noticed by Mr. Austen,[2] is going on eastward of Hurst Castle.

The drift, which spreads over the whole of the New Forest, is not very interesting. No elephants' tusks, or elks' horns, so far as I know, have ever been discovered. A few species of *Terebratula* and *Pecten*, some flint knives, and the *os inominatum*, of probably *Bos longifrons*, mentioned farther on, are the only things at present found. Still, in one way, it is most interesting, as completely disproving the Chroniclers' accounts that, before its afforestation by the Conqueror, the district of the Forest was so fertile. The fact is a sheer impossibility. No wheat could ever be grown on this great bed of chalk-gravel, which is varied only by patches of sand.

But nowhere, perhaps, in the world can we see the stratification of the upper portion of the Middle-Eocene better than at Hordle and Barton, as the sea serves to keep the different strata exposed. The beds dip easterly with a fall of about one in a hundred, though, at the extreme west, at High Cliff, it is much less, and here and there in some few places they lie almost horizontally.[3] At Hordle they seem to have been deposited in a river of a very uniform depth. There is but one single fault in the whole series, just under Mead End, where all the beds have alike suffered. Here and there, however, they are deposited with an undulating line; and here and there, too, a rippled surface occurs, caused by the action of small waves. The river appears to have varied very much in the amount and force of its stream, as some of the beds, where the shells are less frequent, have been deposited very rapidly, whilst others, where the organic remains are more abundant, have been laid on very slowly and in very still water.[4]

It will be impossible to examine all the beds. One or two, however, may be mentioned. And since the beds rise at the east we will begin from Milford. First of all, at Mineway, there runs a remarkable band of fine sand, the "Middle Marine Bed," discovered some twenty-five years ago, by Mr. Edwards, and subsequently successfully worked by Mr. Higgins. It is seldom, however, exposed for more than a few yards; but that is sufficient to show, that after the elevation of the beds beneath they once more subsided, and the sea came over them again, and after that they were once again elevated.

Just below Hordle House rises the "Crocodile Bed," running out of the cliff about three hundred yards from Beckton Bunny. The lowest part of it teems with fish-scales, teeth, crocodile plates, ophidian vertebrae, seed vessels, and other vegetable matter, very often mixed in a coprolitic bed, just beneath a band of tough clay, the specimens being more frequent to the east than the west. The accompanying section (I.) will, perhaps, not only serve to show the situation of the bed, but also those above and below. My measurements will be found to differ slightly from Sir Charles Lyell's[5] and Dr. Wright's;[6] but this is owing to their having been taken in different places.

Immediately under the "Leaf Bed," which, as seen in the opposite section, rises from the shore to the west of Hordle House, comes the lowest bed of the Lower Freshwater Series, formed of blue sandy clay sixteen feet in thickness, from whence Mr. Falconer obtained so many of his mammalian remains.[7]

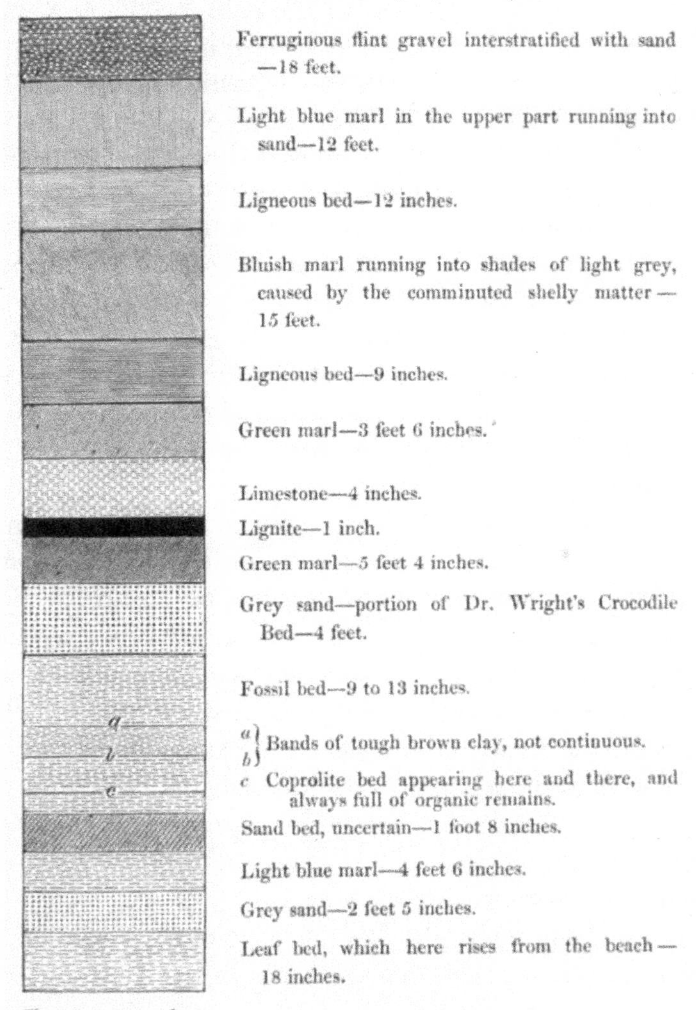

Section I. *of Hordle Cliff, a little to the west of Hordle House.—The beds here incline at an angle of 5°.*

It is a bed, however, which is seldom open, and can be worked only at particular tides. It may easily be recognized as lying between the Leaf Bed and the well-marked Lignite Bed, which shows the first traces of salt-water, and where, in the lower portion, *Neritina concava* may be abundantly found. This last bed may be well seen at Beckton Bunny (section II.). The lignite, however, though it will give a good deal of heat, will not blaze. Locally it is sometimes used for making black paint.

SECTION II. *of Beckton Cliff immediately to the west of the Bunny.*

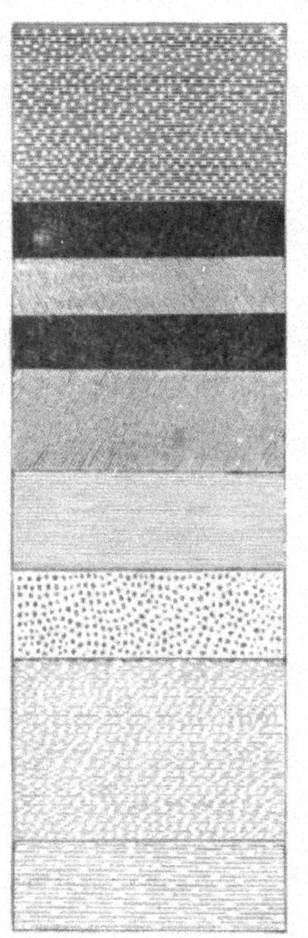

Flint gravel—scarcely more than 3 or 4 feet, with an uncertain band of white sand.

Lignite—3 inches.

Brown clay—3 inches.

Lignite—3 inches.

Marl and sand—2 feet 2 inches.

Ligneous bed, containing shells much broken—8 inches.

Grey sand—2 feet 4 inches.

Orange-coloured sand, with very few fossils at this point, though plenty eastward—15 feet 9 inches.

Olive bed. Fossils abundant—27 feet 3 inches.

The present sea-shore.

Passing on to Beckton Bunny we reach the first true bed of the Lower Marine Formation, which rises a little eastward of that ravine. I have distinguished it as the Olive Bed, from the abundance of specimens of *Oliva Branderi*, forming the equivalent to number eighteen in Dr. Wright's arrangement, and which, when worked, emits a strong smell of sulphur.

Immediately under the Olive Bed, as seen in the opposite section (II.), taken immediately on the west side of the Bunny, rises grey sand, seventeen feet and half in thickness, possessing only a few casts of shells. The next bed, however, composed also of grey sand, rising about three hundred yards farther on, is, perhaps, the richest in the whole of this Marine series, and its shells the best preserved. It may at once be recognised by the profusion of *Chama squamosa*, from which it has been called the Chama Bed. Specimens of *Arca Branderi* and *Solen gracilis* may be found here as perfect as on the day they were deposited.

A little farther on, nearly under the Gangway, rises the Barton clay, encrusted with *Crassatella sulcata*.[8] And here, on looking at the cliff, we may notice how all the beds, as they rise westward, gradually lose their clayey character, and run into sand, which will account for this part of the cliff foundering so fast. The water percolates through the sand down to the Barton Beds, and the loose mass above is thus launched into the sea.

Below the Barton Coastguard Station rises another bed of green clay, containing sharks' teeth and the bones of fish. About a mile farther on, the High Cliff Beds emerge rich with *Cassis ambigua* and *Cassidaria nodosa*. And below them, seen in the channel of the stream flowing through Chewton Bunny, rises a bed of bright metallic-looking, green clay, the *Nummulina Prestwichiana* Bed of Mr. Fisher, containing sharks' teeth and some few shells. Beyond, a little to the west of High Cliff Castle, occurs the well-marked Pebble Bed, the commencement of the Bracklesham Series, containing rolled chalk flints, and casts of shells. Next follow grey sands full of fossil wood and vegetable matter, marked by' a course of oxydized ironstone-septaria.

Then succeeds another Pebble Bed, and lastly appear the grey Bracklesham Sands.[9]

We have thus gone through the principal beds, both of the Freshwater and Marine Series, as far as they are exposed in this section along the sea-coast. The fluvio-marine beds stretch away eastward as far as Beaulieu and Hythe, but their clays here contain very few shells. On the other hand, the Bracklesham Beds, trend away northward towards Stony-Cross, appearing in the valley, and cropping out again on the other side of the Southampton Water.

Some few words must be said about them. The highest beds, known as the Hunting Bridge Beds, occur in Copse St. Leonards, not far from the Fritham Road.[10] In a descending order, separated by thirty or forty feet of unfossiliferous clays, come the Shepherd's Gutter Beds, to be found about half-a-mile lower down the King's Gairn Brook; and below them, again, separated by forty or fifty feet of unfossiliferous clays, and situated somewhat more than a mile lower down the same stream, rise the Brook Beds. Still farther down, too, from some shells very lately discovered at Cadenham, it is supposed that the *Cerithium* Bed of Stubbington and Bracklesham Bay will be found, but this is not yet ascertained.

The Hunting Bridge Beds I have never examined, but subjoin their measurements, as also their most typical shells,[11] and must here content myself to give a general description of the Shepherd's Gutter and Brook Beds. The former, the equivalent to the *Nummulina* Bed at Stubbington, Bracklesham, and White-Cliff Bay, is so called from a small stream at the foot of Bramble Hill Wood, about a mile due north of the King's Gairn Brook. The measurements are as follow :— (1) Gravel from one to five feet; (2) light-coloured clay, with a few fossils sparingly distributed, five to six feet; (3) *Turritella carinifera* bed, one foot and a half; (4) fossil bed, characterized by *Conus deperditus*, and the abundance of *Pecten corneus* within a few inches of the bottom, one foot and a half.

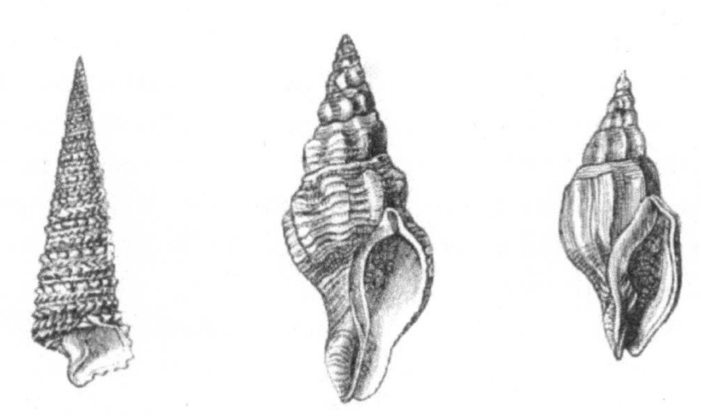

Shells from the Shepherd's Gutter Beds

It is worth noticing that these, like all the Bracklesham beds, roll. In a pit which Mr. Keeping and myself dug we found there had been a regular displacement of the gravel, and that the beds rose at an angle of thirty degrees, whilst the fossil bed was three feet lower on one side than the other of the pit. In another, after cutting through a foot of gravel, in which we found the *os inominatum*, of probably *Bos longifrons*,[12] and a bed of sandy clay about two feet in thickness, we came upon a deposit of gravel about four inches thick, lying in the depressions of the stiff brown clay which succeeded, and in which still remained roots and vegetable matter. Thus we can plainly see that, after the clay had been deposited, vegetable, and perhaps animal, life flourished. Then came the gravel, carrying all before it, and in its turn, too, was nearly swept away, and only left here and there in a few scattered patches.

Perhaps, nothing is so startling as this insecurity of life. As was the Past so will be the Future, guided, though, always by that Law, which at every step still rises, moving in no circle, but out of ruin bringing order, and from Death, Life.

The Brook Beds I can best describe for the general reader by an account of a pit which Mr. Keeping and myself made. It was sunk about 20 feet from the King's Gairn Brook, and measured about 6 yards long by 4 broad. We first cut through a loamy sand, measuring 8 feet, and then came upon 19 inches of gravel, where at the base stretched the half fossilized trunk of an oak, and a thick drift of leaves mixed with black peaty matter, the remains of some primæval forest. Three feet of light-coloured clay, unfossiliferous, succeeded; and then came the *Corbula* Bed, with its myriads of *Corbula pisum*, massed together, nearly all pierced by their enemies, the *Murices*. Stiff light-coloured clay, measuring 18 inches, followed, revealing some of the shells, which were to be found so plentiful in the next stratum. Here, at the *Pleurotoma attenuata* Bed, our harvest commenced, and since Mr. Keeping has worked these beds, no spot has ever yielded such rich results. Every stroke of the pick showed the pearl and opal-shaded colours of the nautilus, and the rich chestnut glaze of the *Pecten corneus*, whilst at the bottom lay the great thick-shelled *Carditæ planicostæ*. Inside one of these were enclosed two most lovely specimens of *Calyptræa trochiformis*. Mr. Keeping here, too, found a young specimen of *Natica cepacea* (?), and I had the good fortune to turn up the largest *Pleurotoma attenuata* ever yet discovered, measuring 4½ inches in length, and 3¼ inches in circumference round the thickest whorl.

We were now down no less than 8 feet. And at this stage the water from the brook, which had been threatening, began to burst in upon us from the north side. We, however, with intervals of bailing, still pushed on till we reached the next bed of pale clay, measuring from 7 to 8 inches, containing *Cassidariæ* highly pyritised, and sharks' teeth, amongst which Mr. Keeping discovered an enormous spine, measuring at least 10 inches in length, but we were unable to take it out perfect. The water had all this time been gaining upon us, in spite of our continuous efforts to bail it with buckets. We, however, succeeded in making the *Voluta horrida* bed, which seemed, at this spot, literally teeming with shells. Each spitful, too, showed specimens of fruit, earbones, fish-palates, drift-wood, and those nodular concretions which had gathered round some berry or coral.[13]

At this point, the water, which was now pouring through the side in a complete stream, and a rumbling noise, showed danger was imminent. Hastily picking up our tools and fossils we retreated. In a moment a mass of clay began to move, and two or three tons, completely burying our bed, fell where we had stood. Founder after founder kept succeeding, driving the water up to higher levels. We procured assistance, but precious time was lost. Night began to fall, and we were obliged to leave unworked one of the richest spots which, in these beds, may, perhaps, ever be met.

As it was, we found no less than sixty-one species, including in all 230 good cabinet specimens, which, considering the small size of the pit, and our limited time, and the great disadvantages under which we worked, well showed the richness of these beds.

Merely, however, collecting fossils for collecting's sake is useless. The aim of geology is to enable us to understand how this world was made—how form followed form, how type after type took life and then passed away, and the higher organization ever succeeded the lower. The Middle-Eocene ought to be to us particularly interesting, separating us, on the one hand, from those monsters which had filled the previous Age, and, on the other, presenting the first appearances of those higher mammals which should serve the future wants of man. The pterodactyle no longer darkened the air. The iguanodon now slept in its grave of chalk. A new earth, covered with new types and new forms, had appeared. It is a strange sight which the Hordle Cliffs unveil. Here, beneath a sun fiercer than in our tropics, the crocodile basked in its reed beds. Here the alligator crimsoned the stream, as he struck his jaws into his victim; whilst the slow tryonyx paddled through the waves, and laid its eggs on the sand, where its plates are now bedded.

The very rushes, which grew on the river banks, lie caked together, with the teeth of the rats which harboured in them. The pine-cones still, too, lie there, their surfaces scarcely more abraded than when they dropped from the tree into the tepid waters. Along the muddy river shore browsed the paloplothere, whilst his mate crushed through the jungle of club-mosses. Groves of palms stood inland, or fringed the banks, swarming with land-snakes. Birds waded in the shallows. But no human voice sounded: nothing was to be heard but the screaming of the river-fowl, and the deep bellow of the tapir-shaped palæothere, and the wolf-like bark of the hyænodon.

This description is no mere fancy, but taken from the remains actually discovered in the Hordle Cliffs. I have had no need to borrow from the fossils of the Headon and Binstead Beds, or the caves of Montmartre. On these cliffs, too, is scored the history of the past. Here lie the little *Nuculæ*, still crimson and pink as when they first settled down through the water into their bed of sand; and teeth of dichodons still bright with enamel. The struggle of life raged as fiercely then as now. And the pierced skull of the palæothere still tells where it received its death-wound from its foe the crocodile.

But other things do they reveal. They plainly show, as was, I believe, first suggested by Mr. Searles Wood, that in the Middle-Eocene period Europe and America were connected. The pachyderms of Hordle are allied to the tapirs of the New World. The same alligators still swim in the warm rivers of Florida: and the same type of sauroid fish, whose scales spangle the Freshwater Beds, is now only found in the West.

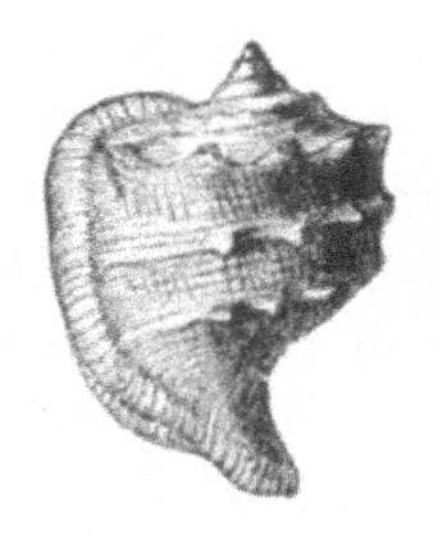

Shells from the Brook Beds.

Footnotes

1. See chap, v., pp. 57, 58. It is just possible that by his "τὰς πλησίον νήσους," Diodorus may mean the Shingle Islands, which we have described in chapter xiv. p. 151, and whose sudden appearance and disappearance would lead to the most extravagant reports.
2. "On the Newer Deposits of the Sussex Coast:" *Geological Journal*, vol. xiii. pp. 64, 65.
3. In the coast-map at p. 148, the principal beds are marked, so that, I trust, there will be no difficulty in finding them.
4. For the direction of the river from east to west, see a paper "On the Discovery of an Alligator and several New Mammalia in Hordwell Cliff," by Searles Wood, F.G.S.: *London Geological Journal*, No. 1., pp. 6, 7.
5. "The Freshwater Strata of Hordwell Cliff, Beacon Cliff, and Barton Cliff:" *Transactions of the Geological Society*, second series, vol. ii., p. 287.
6. "Stratigraphical Account of the Section of Hordwell, Beckton, and Barton Cliffs:" *The Annals and Magazine of Natural History*, June, 1851. In making these measurements I was very greatly assisted by the Rev. W. Fox, who was most untiring to ensure accuracy.

7. See the *Geological Journal*, vol. iv., p. 17; as also, Professor Owen's *Monograph* on "The Fossil Reptilia of the London Clay," published by the Palæontographical Society, 1850, p. 48.
8. Some of the most characteristic shells in this bed may perhaps be mentioned:—

Pleurotoma exorta. *Sol.*	Scalaria reticulata. *Sow.*
Terebellum fusiforme. *Lam.*	Scalaria semicostata. *Sow.*
Murex minax. *Sol.*	Littorina sulcata. *Pilk.*
Murex asper. *Sol.*	Solarium plicatum. *Lam.*
Murex bispinosus. *Sow.*	Hipponyx squamiformis. *Lam.*
Typhis pungeus. *Sol.*	Fusus porrectus. *Sol.*
Voluta ambigua. *Sol.*	Fusus errans. *Sol.*
Voluta costata. *Sol.*	Fusus longævus. *Sol.*
Voluta luctatrix. *Sol.*	Bulla constricta. *Sow.*
Dentalium striatum *Sow.*	Bulla elliptica. *Desh.*

I scarcely need, I hope, refer the reader either to Mr. Edwards' *Monograph on the Eocene Mollusca*, 1849, 1852, 1854, 1856, or to Mr. Searles Wood's *Monograph* on the same subject, both in course of publication by the Palæontographical Society. There is an excellent table of the Barton shells, by Mr. Prestwich, in the *Geological Journal*, vol. xiii. pp. 118-126.

9. For the High Cliff Beds, see Mr. Fisher's paper on the Bracklesham Sands of the Isle of Wight Basin, in the *Proceedings of the Geological Society*, May, 1862, pp. 86-91, whose divisions are here followed.
10. All these beds are shown in the large map by the word "Fossils," there not being space enough to particularize each bed.
11. These beds were discovered by Mr. Fisher in 1861, and for the following measurements I am indebted to Mr. Keeping. We find, about one hundred yards in a south-eastward direction from the point where the footpath from Brook to Fritham crosses the stream, (1) the Coral Bed, the equivalent of that at Stubbington, full of crushed *Dentalia* and *Serpulæ*, six inches. (2) Sandy light blue clay, with very few fossils, seven feet. (3) Verdigris-green and slate-coloured clay, characterized near the top by a new species of *Dentalium*, *Serpulorbis Morchii* (?), and *Spondylus rarispina*. The other typical shells are *Voluta Maga*, several species of *Arca* and *Corbula gallica*, five feet. It is in this bed that large roots of trees and ferns are found.
No persons, however, I should suppose, would think of examining any of these beds without first consulting Mr. Fisher's most valuable paper on the Bracklesham Beds in the *Proceedings of the Geological Society*, May, 1862. And I should further most strongly advise them, if they wish to become practically acquainted with the beds, to procure the assistance of Mr. Keeping, of Freshwater, in the Isle of Wight.
I may here further mention that a well is at the present moment being sunk at Emery Down, and which, as I learn from Mr. Keeping, gives the following interesting measurements:—(1) Beds of marl, containing *Voluta geminata*, discovered forty years ago, at Cutwalk Hill, by Sir Charles Lyell, and now re-discovered, and a small *Marginella*, seven feet. (2) Bed of bluish sandy clay, which becomes, when weathered, excessively brown. This bed, very rich in fossils, which are in a good state of preservation, is equivalent

to what is now called the Middle Marine Bed, at Hordle and Brockenhurst, sixteen to nineteen feet. (3) Hordle Freshwater Beds, containing two species of *Potanomya*, and comminuted shells, fifteen feet. (4) Upper Bagshot Sands, measuring, as far as the workmen have gone, twenty feet, and below which lies the water at the top of the clay. The important point to be noticed is the extreme thinning out of the Hordle Freshwater Beds, which, from the depth of two hundred and fifty feet at Barton have here shrunk to fifteen. Mr. Prestwich has suggested that these beds, as they advance in a north-easterly direction, become more marine, which seems here to be confirmed.

12. I say probably, for Professor Owen, who examined the specimen, states that it is of a bovine animal of the size of *Bos longifrons*, but does not yield characters for an exact specific identification. I may here add that the celt mentioned at p. 207, foot-note, is hardly satisfactory.

13. I had intended to have accompanied this description with a group of some of the best fossils from this pit, including the fruit, fish-spines, and palates, and the large *Pleurotoma attenuata*. It was, in fact, commenced by the artist. But the specimens were obliged to be so greatly reduced, that the drawing gave no complete idea of their form and beauty, and would only have confused the reader. I have, therefore, contented myself with figuring at p. 249, in its matrix of clay, the rare *Natica cepacea* (?), which has passed into Mr. Edwards' fine collection, and who has kindly allowed me the use of it, with the characteristic *Cassidaria nodosa*, and a lovely *Calyptrcea trochiformis*, found, as mentioned, inside a *Cardita*. At p. 244, the specimens given from the Shepherd's Gutter Beds are *Cerithium triliuum* (Edw. MS.), *Voluta uniplicata*, and, in the centre, a shell, showing oblique folds on the *columella*, which Mr. Edwards thinks may be identical with *Fusus incertus* of Deshayes.

CHAPTER XXI.

THE BOTANY.—THE FLOWERING PLANTS AND FERNS.

Barrow's Moor Wood

Closely connected with the geology of the Forest are its flowers. And though mere geology could not tell us the whole Flora of a district, yet we might always be able, by its help and that of the latitude, to give the typical plants. Close to the chalk, the Forest possesses none of the chalk flowers. No bee-orchis or its congeners, although so common on all the neighbouring Wiltshire downs, bloom. No travellers'-joy trails amongst its thickets, although every hedge in Dorsetshire, just across the Avon, is clothed in the autumn with its white fleece of seeds. No yellow bird's-nest (*Monotropa Hypopitys*) shades itself under its beeches, though growing only a few miles distant on the chalk.

Still, here there are some contradictions. The chalk-loving yew appears to be indigenous. Several plants which we might reasonably expect, as herb-Paris, the bird-nest orchis (*Neottia Nidus-avis*), and the common mezereon (*Daphne Mezereum*), are wanting.

Owing to the want of stiff clay, no hornbeams grow in its woods, except, perhaps, a few in one or two cold "bottoms." No Solomon's seal or lilies of the valley whiten its dells. No meadow-geranium waves its blue flowers on the banks of the Avon.[1]

On the other hand, the plants too truly tell the character of the soil. In the spring the little tormentil shows its bright blossoms, and the petty-whin grows side by side with the furze, and the sweet mock-myrtle throws its shadow over the streams. In the summer and autumn the blue sheep's-bit scabious and the golden-rod bloom, with the three heathers. In the bogs the round-leaved sundew is pearled with wet, and not far from it the cotton-grass waves its white down, and the asphodel rears its golden spike.

These are the commonest flowers of the Forest, and grow everywhere over its moors. In its dykes and marshes, the common frog-bit and the marsh-pimpernel spring up in every direction. The buckbean, too, brightens every pool on the south side, and is so common near the Avon that many of the fields are called "the buckbean mead," whilst in the northern parts it is known as "the fringed water-lily."

Very rich is the Forest in all these bog-plants. In Hinchelsea and Wilverley Bottoms grow the water-pimpernel (*Samolus*

Valerandi), the lesser bladder-wort, and the bur-reed (*Sparganium natans*) floating on the water. Here, too, perhaps, the easternmost station known for it, blossoms the butterwort (*Pinguicula Lusitanica*), with its pale delicate flowers. In the autumn, also, the open turf grounds round Wootton are blue with the Calathian violet (*Gentiana Pneumonanthe*); whilst its little bright congener (*Cicendia filiformis*) blossoms in all the damp places.

Owing, also, to the presence of iron, the Forest possesses no less than seventeen or eighteen carices. The little thyme-leaved flax, too (*Radiola millegrana*), grows in all the moist, sandy dells.

From this general view it will be seen that the true Forest plants are not so much "sylvestral" as "ericetal," and "paludal," and "uliginal." Besides these groups, however, the Flora of the district further divides itself into the "littoral plants" along the sea-shores and estuaries, and the "pascual" flowers of the valley of the Avon. In the former division, owing to the want of rocks, no *Statice spathulata* grows on its sea-board. No true samphire (*Crithmum maritimum*) blossoms. The beautiful maiden's-hair fern, once so plentiful on the neighbouring coast of the Isle of Wight, is also from the same cause wanting.

Still, great beauty blooms on the Forest streams and shores. In the latter part of the summer, the mudbanks of the Beaulieu river are perfectly purple with the sea-aster, whilst the sea-lavender waves its bright blue crest among the reed-beds washed over by every tide.

The valley of the Avon is characterized, as may be expected, by the commoner species, which are to be found in such situations. Here, and in the adjoining cultivated parts, which once were more or less a part of the Forest, we find the soap-wort (*Saponaria officinalis*) and the thorn-apple (*Datura Stramonium*), and those colonists which always harbour close to the dwellings of man. Other considerations remain. The situation and climate of the

New Forest, of course, have a great effect on its plants.[2] The two myrtles and the sweet-bay grow under the cliffs of Eagleshurst, close to the Solent, unhurt by the hardest frosts. The grapes ripen on the cottage-walls of Beaulieu nearly as early as in Devonshire. I have seen the coltsfoot in full blossom, near Hythe, on the 27th of February; and the blackthorn flowers at Wootton on the 3rd of April.

The area of the New Forest comes under Watson's Subprovince of the Mid-Channel, on the Southern belt of his Inferagrarian zone. Its position lies exactly half-way between his Germanic and Atlantic types. The former shows itself by *Dianthus Armeria*, and *Pulicaria vulgaris*, growing near Marchwood and Bisterne. The latter by such examples as *Cotyledon umbilicus*, *Pinguicula Lusitanica*, *Briza minor*, and *Agrostis setacea*. The "British" and "English" types are, of course, plentifully represented.[3]

Looking, too, at the trees and shrubs which are indigenous, we shall find them also eminently characteristic. In spite of what Cæsar says, the beech is certainly a native, pushing out in places even the oak. The holly, too, grows everywhere in massy clumps. In the spring, the wild crab (*Pyrus Malus*) crimsons the thickets of Brockenhurst, in the autumn the maple. The butcher's broom stands at the foot of each beech, and the ivy twines its great coil round each oak, and the mistletoe finds its home on the white poplar.

After all, the trees, and not the flowers, give its character to the New Forest. In the spring, all its woods are dappled with lights and shades, with the amber of the oak and the delicate soft-gleaming green of the birch and beech. In the autumn, the spindle-tree (*Euonymus Europæus*) in the Wootton copses is hung with its rosy gems; and the trenches of Castle Malwood are strewed with the silver leaves of the white-beam.

To return, however, to the plants, let us notice how some particular families seem especially to like the light gravelly soil of the Forest district. Take, for instance, the St. John's-worts, of which we have no less than six, if not more varieties. The common perforated (*Hypericum perforatum*) shines on every dry heath, and the square-stalked (*quadrangulum*) in all the damp boggy places. The tutsan (*Androsæmum*) is so common round Wootton that it is known to all the children as "touchen leaves," evidently only a corruption of its name; and its berries are believed throughout the Forest to be stained with the blood of the Danes. The rarer large-flowered (*calycinum*) grows, though not, I am afraid, truly wild, in some of the thickets round Sway. In all the ponds, the marsh (*elodes*) springs up, whilst the creeping (*humifusum*) trails its blossoms over the turf of the Forest lanes, and the small (*pulchrum*) shows its orange-tipped flowers amongst the brambles and bushes.

Take, again, the large family of the ferns, of which seventeen species are distributed throughout the Forest. First and foremost, of course, stands the royal fern (*Osmunda regalis*), which may be found from the sea-board to Fordingbridge, rearing its stem in some places six feet high, and covering in patches on the southern border, as at Beckley, nearly a quarter of an acre. It grows in Chewton Glen, in all the lanes in the neighbourhood, on Ashley Common, close to the Osmanby Ford River, and rears its golden-brown pannicles in the boggy thickets near Rufus's Stone. But before it, in beauty, stands the lady-fern, with its delicate fronds and its tender green, growing in the open spaces of the beech woods, as at Stonehard and Puckpits, and bending over the Forest streams in large leafy clumps. Then, too, in all the large woods grows the sweet-scented mountain fern (*Lastrea Oreopteris*); and on every bank the hart's-tongue spreads its broad ribbon-like leaves, and the fertile fronds of the hardfern spring up feathery and light, whilst from the old oaks the common polypody droops with its dark green tresses. The common maiden-hair (*Asplenium Trichomanes*), too, hangs on the walls and Forest banks; and on Alice Lisle's tomb, at Ellingham, the rue-leaved spleenwort is green throughout the whole year. On Breamore churchyard wall and Ringwood bridges grows the common scale-fern, whilst in the meadows of the Avon springs the adder's-tongue's green spear.

Nor must we forget the brake, common though it be, for this it is which gives the Forest so much of its character, clothing it with green in the spring; and when the heather is withered, and the furze, too, decayed, making every holt and hollow golden.[4]

And now for some other plants, without reference to their species, but simply to their beauty. On Ashley Common and the neighbouring grass-fields grows the moth-mullein (*Verbascum Blattaria*), dropping its yellow flowers, as they one by one expand. In the neighbouring pools, as far as Wootton, the blossoms of the great spearwort (*Ranunculus Lingua*) gleam among the reeds. There, also, the narrow-leaved lungwort (*Pulmonaria angustifolia*), with its leaves both plain and spotted, opens its blue and crimson flowers so bright, that they are known to all the children as the "snake flower," and gathered by handfuls mixed with the spotted orchis. And the ladies' tresses, too (*Spiranthes autumnalis*), shows its delicate brown braid on every dry field on the southern border.

Besides these, the feathered pink (*Dianthus plumarius*) blooms on the cloister-walls at Beaulieu; and the Deptford pink (*Dianthus Armaria*) in the valley of the Avon at Hucklebrook, near Ibbesley. The bastard-balm (*Melittis Melissophyllum*) flaunts its white and purple blossoms over the banks of Wootton plantation, whilst at Oakley and Knyghtwood the red gladiolus crimsons the green beds of fern.

Briefly, let me say that, as is the Forest soil, so are its plants. Nature ever makes some compensations. The barrenest places she ever clothes with beauty. If corn will not grow, she will give man something better. In the great woods the columbines and tutsan shine in the spring with their blue and yellow blossoms, and the wood-sorrel nestles its white flowers among the mossy roots of the oaks. In the more open spaces the foxgloves overtop the brake, and in the grassy spots the eyebright waves its white-grey crest; and not far off are sure to gleam faint crimson patches of the marsh-pimpernel, half hid in moss; whilst the swamps are fringed with the coral of the sundew.

The King's Gairn Brook (Another View).

Footnotes

1. In one place only in the Forest, on some waste ground at Alum Green, have I seen this plant.
2. On this point see what Bromfield observes in his Introduction to the *Flora Vectensis*, p. xxvi.
3. In Appendix II. I have given a list of all the characteristic

plants of the New Forest to assist the collector; and, I trust, comprehensive enough for the botanist to make generalizations.
4. Besides these we have all over the Forest *Lastrea Filix-mas*, and *dilatata*, and *Asplenium adiantum nigrum*, and *Polystichum angulare*, with its varieties, *angustatum* and *aculeatum*, found near Fordingbridge. My friend, Mr. Rake, who discovered *angustatum*, found also, in February, 1856, near Fordingbridge, *Lastrea spinulosa*, but it has never since been seen in the locality.

CHAPTER XXII.

THE ORNITHOLOGY.

The Heronry at Vinney Ridge.

To describe the Fauna of the Forest is beyond the purpose of this book, and would, beside, require a life-time to properly accomplish. I can only here deal with the ornithology as I have with the botany. I do not know either that the general reader will lose anything by the treatment. A scientific knowledge is not so much needed as, first of all, a sympathy with nature, and a love for all her forms of beauty. The great object in life is not to know, but to feel. But, before we speak of the birds, let us correct some errors which are so common with regard to the animals. It is quite a mistake to talk of wild boars or wild ponies roaming over the Forest. There is not now an animal here without an owner. The wild boars introduced by Charles I., and others brought over some fifty years ago, are seen only in their tame descendants—sandy-coloured, or "badger-pied," as they are called, which are turned out into the Forest during the pannage months.[1]

So, too, the Forest ponies never run wild, except in the sense of being unbroken. Lath-legged, small-bodied, and heavy-headed, but strong and hardy, living on nothing in the winter but the furze, they are commonly said, without the slightest ground, to be descendants of the Spanish horses which swam ashore from the disabled ships of the Armada.

And now for the ornithology. The thick woods, the lonely moors and holts, attract the birds of prey; the streams and marshes the waders; whilst the estuaries of the Beaulieu, and Lymington, and Christchurch rivers, and the Solent, afford a shelter in winter to the geese and ducks driven from the north.

Again, too, the peculiar mildness of the climate has its effect on the birds as well as the plants. The martin and the swallow come early in March and stay till the end of November; that is to say, remain full three-quarters of the year. I have heard, too, the cuckoo as early as April 11th and as late as July the 12th. The warblers, whose arrival depends so much on the south-east winds, may not come earlier than in other parts of England. They certainly, however, in the southern and more cultivated parts, where food is plentiful, stay here later than in the Midland Counties; and I have heard the whitethroat singing, as on a spring day, in the middle of October.

We will begin with the birds of prey. Gilpin (vol. ii. p. 294) mentions a pair of golden eagles, which, for many years, at times frequented King's Wood, and a single specimen, killed near Ashley Lodge. These, however, with the exception of one shot some twenty years ago over Christchurch Harbour, are the last instances of a bird, which can now be seldom seen except in the north of Scotland. Yarrell,[2] too, notices that the sea eagle (*Aquila albicilla*) is sometimes a visitor in the district, but though I have been down under the Hordle and Barton Cliffs, day after day, for often six months together, I have never seen a specimen. Still it sometimes occurs in the winter, and is mistaken for its rarer ally; and the Eagle Tree at the extreme west end of Vinney Ridge still commemorates where one was shot, some fifty years ago, by a Forest-keeper. The osprey, however (*Falco haliæëtus*), still frequents the coast in the autumn, and still circles over Christchurch Harbour fishing for his prey, where, as Yarrell mentions, he is well known as the "grey-mullet hawk," on account of his fondness for that fish.

The Peregrine Falcon (*Falco peregrinus*), which breeds on the high Culver Cliffs of the Isle of Wight, and in the Lulworth Rocks, is in the summer a regular visitor, and scours the whole country. No year goes by without some half-dozen or more being killed.

Its congener the hobby (*Falco subbuteo*), known in the Forest as "the van-winged hawk," comes about the same time as the honey-buzzard, building in the old, deserted nests of crows and magpies, and even, as in one case, to my knowledge, in that of the honey-buzzard. The bird, however, is becoming scarce. For several years I have known a pair or two build in Buckhill Wood, of which a sketch is given at the end of this chapter, but last year none came. It lays generally about the beginning of June, though I have received its eggs as late as July 12th. Yarrell says that their number is three or four; but, with Mr. Hoy,[3] I have never known the bird lay more than three, and very often only two.

The goshawk (*Falco palumbarius*) and the rough-legged buzzard (*Falco lagopus*) are very rarely seen; but, I fear, the kite, although so plentiful in Gilpin's time, has nearly deserted this, like all other districts. Once, and once only, has it been seen by Mr. Farren. The honey-buzzard, however (*Falco apivorus*), comes regularly over from Germany about the end of May, attracted, in some measure, perhaps, by its favourite food, the larvae of wasps and bees, but chiefly by the wide range of the woods. At Mark Ash and Puckpits I have frequently, for an hour together, watched a couple, sailing with their wings outspread, allowing the wind, on a boisterous day, to catch them, till it almost veered them over; just circling round the tops of the beeches, sometimes even "tumbling," like a pigeon, and answering each other with their sharp, short cry, prolonged every now and then into a melancholy wail. Its favourite breeding stations are amongst the tall beech-woods round Lyndhurst, in Mark Ash, and Gibbs Hill, Puckpits, Coalmeer, Prior's Acre, and the oaks of Bentley and Sloden. The nest is always placed in the old one of a crow, or even the common buzzard, whose young by that time have flown, and sometimes made on the top of a squirrel's "cage," the birds contenting themselves with only re-shaping it, and lining the inside with fresh green leaves. The fact of a squirrel's "cage" being used will account for the nest being sometimes found so low, and on a comparatively small tree. No rule can therefore be laid down as to its position. I have known the bird build in very different situations. Mr. Rake found its nest in Sloden, on the forked bough of a low oak, not thirty feet from the ground. In 1860 a pair built, not very much higher, in the overhanging branch of a beech in Puckpits; and, in the same year, another pair reared their young on the top of a fir in Holmy Ridge Hill. And in 1861 and 1862, I knew of two nests, not fifty yards apart, in Mark Ash, each placed nearly at the top of the very tallest beeches in the wood, at least seventy or eighty feet from the ground. As so little appears to be known about its breeding habits, I may as well add a few more words. It seldom arrives till the beginning of June, when the leaves are thick on the trees, and immediately commences its nest, for which purpose it seems only to come, as it immediately departs when the young birds can fly. Pairs have been known, however, not to lay till the end of July; and, I am assured by one of the Forest keepers, not sometimes till even the beginning of August; but these are, doubtless, cases where the birds have been robbed of their first eggs. It differs from the common buzzard in not flying away when disturbed during incubation, but merely skimming round the top of the tree in small circles, uttering its short, shrill cry, sometimes both male and female perching on the branch of a neighbouring tree, and remaining undisturbed by shouts or cries, whilst the nest is being reached. At these times a kind of stupidity seizes the bird. It has, to my knowledge, on several occasions, remained in the nest till a boy has touched its feathers, and returned as soon as he left.

As a further illustration, I may add, that in one of the nests before mentioned, in Mark Ash (June 7th, 1862), was only one egg, which was taken. The birds, however, did not forsake, and another, which was also taken, was laid on the third day. Even then the birds did not desert, but after the interval of two more days laid a third egg, about one-half smaller than usual, and in shape somewhat resembling a peregrine's.

On another occasion, June 11th, 1859, a pair bred in a high beech in Coalmeer Wood, near Stoney Cross, and though fired at more than once did not desert. The female, however, was first shot, when the cock, nothing daunted, took his partner's place, and sat on the eggs, and in a day or two afterwards shared her fate. In the nest were two eggs, which, with the exception before mentioned, I have never known exceeded. Those in my collection vary in colouring from the light dull vermilion, which so often characterizes the merlin's eggs, to a deep rich morone, tinted, especially in newly-taken specimens, with a delicate crimson bloom.[4]

A few words more. The birds are not much seen in the day, but generally early in the morning. Whilst the hen bird sits on the eggs, the cock perches close by in some tall thick tree. Perhaps from this very affection for their young arises their seeming stupidity, and the ease with which they are killed. Some years ago a keeper found a nest with two young birds in Bentley Wood, and on purpose to secure them tied them by their legs to a small tree, where the old birds regularly came and fed them. But the strangest fact with regard to their breeding is that before they finally decide upon a nest they will line several with green leaves and small leafy twigs. Lastly, I may add that though I have examined many nests, I have never found any traces of their being, as is related by some writers, lined with wool. If there was any wool it was probably placed there by the bird which had previously inhabited the nest.

The common buzzard (*Falco buteo*) is a resident all through the year in the Forest, and may now and then be seen towering high up in the air, so high that you would not at first notice him, unless you heard his wild scream. It is not, however, nearly so plentiful as formerly. He is a sad coward, and the common crow will not only attack, but defeat him. Once or twice I have seen their battles during the breeding season. The jays, and magpies, too, and even the pewits, will mob him, the latter striking at him almost like a falcon. Its favourite breeding-places are in the Denny and Bratley Woods, Sloden, Birchen Hat, Mark Ash, and Prior's Acre. Several nests are yearly taken, for the bird generally breeds when the bark-strippers are at work in April and May. A series of its eggs, in my collection, taken in the Forest, show every variety of colouring from nearly pure white to richly blotched specimens.

In the breeding-season the birds are excessively destructive. A boy who climbed up to a nest in the spring of 1860 told me that he found no less than two young rabbits, a grey hen, and two thrushes as provision for two nestlings. However, there is always some compensation, for in one which I examined were the skeletons of two snakes and a rat picked to the bone.

The accompanying vignette will, I trust, although the nests are so exactly alike, be of some interest. Whilst the artist was sketching the honey-buzzard's nest, the old bird, the first which I had noticed in 1862, made its appearance and circled round the tree, uttering its peculiar short shrill squeak. This nest, which had been repaired in the previous year, the dead beech-leaves still hanging on to the twigs, was between forty and fifty feet from the ground; whilst that of the common

Common Buzzard's Nest

Honey-Buzzard's Nest.

buzzard, who, whilst sitting, had, a month before, been killed, was upwards of seventy feet, and placed on the very topmost boughs of a beech, on which tree was also the other.

But more important than even the nesting of the honey-buzzard is that of the merlin (*Falco æsalon*), which fact has never yet been, so far as I know, noticed as occurring in the New Forest. In the winter this little hawk is sometimes seen hunting, as it does in Ireland, the snipe, although but few specimens find their way to the bird-stuffer. It lingers on, however, to the summer, but the opportunities then of watching its habits are more rare, as the foliage of the woods is so thick. In 1859 and 1861 Mr. Farren received two nests with three eggs, taken in old pollard hollies growing in the open heath, which in every way corresponded with those of the merlin, being considerably smaller than those of kestrels. Unfortunately, however, he could not procure the parent birds, and the fact of the merlins' nesting remained doubtful. In 1862 he was at last successful, and on May 22nd discovered a nest, placed in the hole of a yew, also containing, like the others, three eggs, from which the male bird was shot. Both the bird and eggs are now in my collection, the latter being somewhat richer and darker in colour than those which I have received from the Orkney and Shetland islands. The important fact, however, to be noticed is that, as Temminck remarks, the birds in a woody country build in trees, whilst in the north of Britain, where there is no timber, they adapt themselves to the country, and lay on the ground.[5]

The marsh and hen-harriers, too, frequent the moors and heaths of the Forest, especially the latter, locally known as the "blue hawks." Some few pairs of these breed here, and in 1859 a nest containing three young birds was found near Picket Post by a woodman, and another in 1862, with three eggs, on Beaulieu Heath. One of the Forest keepers described the fern for some distance round a nest, which he discovered, as completely trodden down by the young birds, and so littered with feathers and dirt that, to use his words, the place had exactly the appearance of a goose-pen. A woodman, too, who in 1860 was set to watch a pair near Ocknell, gave me an interesting account of his seeing the old birds breaking off the young tops of the fern to form their nest. I have never myself been fortunate enough in the Forest to find their nest, but I have often watched a pair on Black Knoll and Beaulieu Heath skimming over the ground, pausing to hover just above the furze, then flying forward for some ten or twenty yards, turning themselves suddenly sideways; and then again, for a minute, poising, kestrel-like, beating each bush, and every now and then going up a little higher in the air, but quickly coming down close over the cover.

Passing from the falcons, let us look at the owls, of which the Forest possesses four, if not more, varieties. The commonest is the tawny (*Strix aluco*), whose hooting fills the woods all through the winter. At Stoney Cross I have repeatedly heard, on a still November night, a pair of them calling to one another at least two miles apart. It not only breeds in holes of trees, but in old crows'-nests, and will often, when its eggs are taken, lay again within a week. The barn owl, strange to say, is not much more abundant than the long-eared (*Strix otus*), which breeds in the old holly-bushes, generally taking some magpie's nest, where it lays three eggs. Rarer still is the short-eared (*Strix brachyotus*), which visits the Forest in November, staying through the winter, and in the day-time rising out of the dry heath and withered fern.[6]

Leaving the owls, let us notice some of the other birds. Many a time, in the cold days of March, have I seen the woodcocks, in the new oak plantations of Wootton, carrying their young under their wing, clutching them up in their large claws. Here, on the ground, they lay their eggs, which are of the same colour as the withered oak-leaves—a dull ochre, spotted and clouded with brown, and are thus easily overlooked. About the same time, or even earlier—in February—the raven will build, or rather used to, in the old woods round Burley. In 1858 the two last nests were taken, the eggs being somewhat smaller than those which I have received from the Orkneys. Another of its breeding stations was in Puckpits, where, however, it has not built for the last four seasons. Formerly the bird was common enough, as the different Ravensnest Woods still show; and old men in the Forest have told me, in direct opposition, however, to what Yarrell says,[7] that when, as boys, taking its eggs, they were obliged to arm themselves with stones and sticks to drive off the parent birds, who fiercely defended their nests with their claws and bills.

Now it is nearly extinct, though a pair may sometimes be seen wherever there is a dead horse or cow in the district.

Then, when the summer comes, and the woods are green and dark, the honey-buzzard skims round the tops of the trees; and the snipe, whose young have not yet left the swamps, goes circling high up in the air, "bleating," as the common people here call the noise of its wings, each time it descends in its waving, wandering flight; whilst out on the open spaces the whinchat, known throughout the Forest, from its cry, as the "furze hacker," jerks itself from one furze branch to another; and flitting along with it fly a pair of Dartford warblers.

And as, too, evening draws down, from the young green fern the goatsucker, the "night-crow" and the "night-hawk" of the district, springs up under your feet, and settles a few yards off, and then flies a little way farther, hoping to lead you from its white marble-veined eggs on the bare ground.

Such scenes can the Forest show to the ornithologist in spring and summer, nor is it less interesting to him in the winter. Here, as he wanders across some moor, flocks of fieldfares and missel-thrushes start out of the hollies, and the ring-ousel skulks off from the yew. A bittern, its neck encircled with a brown frill of feathers, is, perhaps, wading by the stream; and hark! from out of the sky comes the clanging of a wedge-shaped flock of grey-lag geese.

Instead of a chapter a volume might be written upon the ornithology of the New Forest, especially about the winter visitants—the flocks of pochards, and teal, and tufted-ducks, which darken the Avon, and the swans and geese which whiten the Solent. I have stood for hours on the beach at Calshot, and watched the faint cloud in the horizon gradually change into a mass of wings beating with one stroke, or marked string after string of wigeon come splashing down in the mid-channel. Little flocks of ring-dotterels and dunlins flit overhead, their white breasts flashing in the winter sun every time they wheeled round. The shag flies heavily along, close to the water, with his long outstretched neck, melancholy and slow, and the cry of the kittiwake sounds from the mud-flats.

To leave, however, the winter birds, and to pass on to more general observations, let me notice a curious fact about the tree-creeper (*Certhia familiaris*) in the southern parts of the Forest, Here there are large plantations of firs, and consequently but few holes in the trees. To make up for this deficiency, I have twice found the creeper's nest placed inside a squirrel's "cage," showing the same adaptability to circumstances which is met with in the whole animal creation. Here, too, in these thick firs build great numbers of jays; and I have, when climbing up to their nests, more than once seen a squirrel coming out with an egg in its claw or mouth. I should have been inclined to have doubted the fact had I not seen it. The sucked eggs which are so often found must, therefore, be attributed quite as much to the squirrel as the magpie or the jay, who have so long borne the guilt. Of course, too, from the great extent of wood we should expect to find the woodpeckers very plentiful. The common woodpecker, known as the "yaffingale" and "woodnacker," is to be seen darting down every glade. The greater-spotted (*Picus major*) is not unfrequent, and the lesser-spotted (*Picus minor*) in the spring comes out of the woods and frequents the orchards of Burley and Alum Green, boring its hole in the dead boughs.

And here let me notice the tenacity with which the greater-spotted woodpecker, whose nesting habits are not elsewhere in England so well observable, clings to its breeding-place; for I have known it, when its eggs have been taken, to lay again in the same hole, the eggs being, however, smaller. Mr. Farren tells me that he has observed the same fact, which is curious, as its ally, the green woodpecker, is so easily driven away, by even a common starling.

The presence of the great black woodpecker (*Picus martius*) has long been suspected, especially since a specimen has been killed in the Isle of Wight, and a pair have been seen near Christchurch. [8] Mr. Farren, in 1862, was fortunate enough not only to see the bird, but to discover its nest. On the ninth of June, whilst in Pignel Wood, near Brockenhurst, he observed the hen bird fly out of a hole placed about six feet high in a small oak, from which he had earlier in the season taken a green woodpecker's nest. Hiding himself in the bush-wood, he saw, after waiting about half an hour, the hen return, and had no doubts as to its identity. An endeavour, however, to secure her in the hole, with the butterfly-net which he had with him, was unsuccessful. He was afraid to leave the eggs, as some woodmen were working close by, and so lost any other opportunity of making the capture. The eggs, now in my collection, were four in number, one being slightly addled, and are the only specimens ever taken in England. They were laid on the bare rotten wood, the bird finding the hole sufficiently large, as Mr. Farren had widened it when taking the previous eggs. It is, however, remarkable that such a shy bird should have built in such a scattered and thin wood as Pignel, close to a public thoroughfare, and where the woodmen had for some time past been constantly felling timber.

But what gives the Forest so much of its character is the number of herons who have lately established themselves in various parts. You can scarcely go along a stream-side without surprising some one or two, which, as you approach, flap their large slate-coloured wings, and fly off with a rolling, heavy motion, circling in the air as they go. Down at Exbury, at the mouth of the Beaulieu river, they may be seen in companies of threes and fours, wading in the shallows, probing their long bills into the mud and sand; and then, as the tide comes up, making off to the freshwater ponds. They are, however, I am afraid, rather persecuted, as they never long here remain at one breeding station. They first took up their abode in Old Burley Wood, and then removed to Wood Fidley, and subsequently to Denney, and finally to Vinney Ridge. In 1861, fifty pair, at least, must have built in its tall beeches. On a fine early spring morning, a long grey line of them would perch on the neighbouring green of Dame Slough, picking up the twigs of heather and flying off with them to line their great platforms of nests; and then sailing down to the Blackwater stream, in the "bottom" close by, to fish. In the morning and evening, and, in fact, all through the day, one incessant clamour was going on, and under the trees lay great eels, which had fallen from their nests.

Last year the numbers were greatly decreased, the birds having been, perhaps, driven away by the woodcutters and charcoal-burners employed to cut down the surrounding timber. The sketch which stands at the head of this chapter was taken in June—too late in the year to show any of the nests, but several young birds were still hovering round who had not even then quite quitted. A small colony has, too, established itself at Boldrewood, where I trust it will be protected; for few birds possess so much character, and give so much beauty to the landscape.

Before we conclude, let us glance at some other peculiarities of the Forest district, and its effects on its birds. It is not too far westward for the east winds to bring the hoopoe, so common in Sussex. Throughout the summer of 1861, a pair were constantly flying about and hopping on the "Lawn" near Wilverley Forest Lodge. The black redstart (*Sylvia tithys*) and the fire-crest (*Regains ignicapillus*) just skim its borders in their westerly winter migrations. Small flocks of dotterel make it their halting spot for a few days in spring, on their way to their northern breeding-places. In the winter, its mildness brings numbers of siskins, some few bramblings, and the common and even the parrot crossbill, escaping from the frosts of the north.

Other things may be mentioned. The hawfinches do not stay

all the year round, as might be expected, or, at least, only one or two pairs, simply because there are no hornbeams in the Forest, nor gardens to tempt them with their fruits. The chough, too, is seldom seen, its eggs and young being plundered in the Isle of Wight cliffs and the Lulworth rocks. It is now extinct in Sussex, and will soon be in the New Forest. Yet these birds were once so numerous in England, not only damaging the crops, but unthatching the barns and houses, that a special Act of Parliament was passed against them.[9] Twopence for a dozen heads were given. People were, under various penalties, bound to destroy them, and parishes were ordered to keep chough and crow nets in repair.

There is, unfortunately, no other forest in England by which we can make comparisons with the ornithology of the New Forest. In Churchill Babington's excellent synopsis of the birds of Charnwood Forest, we find only one hundred and twenty-five species, but little more than one-half of those in the New Forest. Out of the three hundred and fifty-four British birds the New Forest possesses seventy-two residents, whilst it has had no less than two hundred and thirty killed or observed within its boundaries.[10] With this we must end. I am afraid it is too late to protest against the slaughter of our few remaining birds of prey. The eagle and kite are, to all purpose, extinct, in England, and the peregrine and honey-buzzard will soon share their fate. The sight of a large bird now calls out all the raffish guns of a country-side. Ornithologists have, however, themselves to thank. With some honourable exceptions, I know no one so greedy as a true ornithologist. The botanist does not uproot every new flower which he discovers, but—for he loves them too well—carefully spares some plants to grow and increase; whilst few ornithologists rest content till they see the specimen safe in their cabinets. This, I suppose, must be, from the nature of the study, the case. Still, however, the love for Nature, and the enthusiasm which it gives, must be regarded as a far greater offset. And here let me, for the last time, say that I feel sure that nobody knows anything of the true charms of the country who is ignorant of natural history. With the slightest love and knowledge of it, then every leaf is full of meaning, every pebble a history, every torn branch, gilded with lichens, and silvered with mosses, has its wonders to tell; and you will find life in the dust, and beauty in the commonest weed.

View in Buckhill Wood.

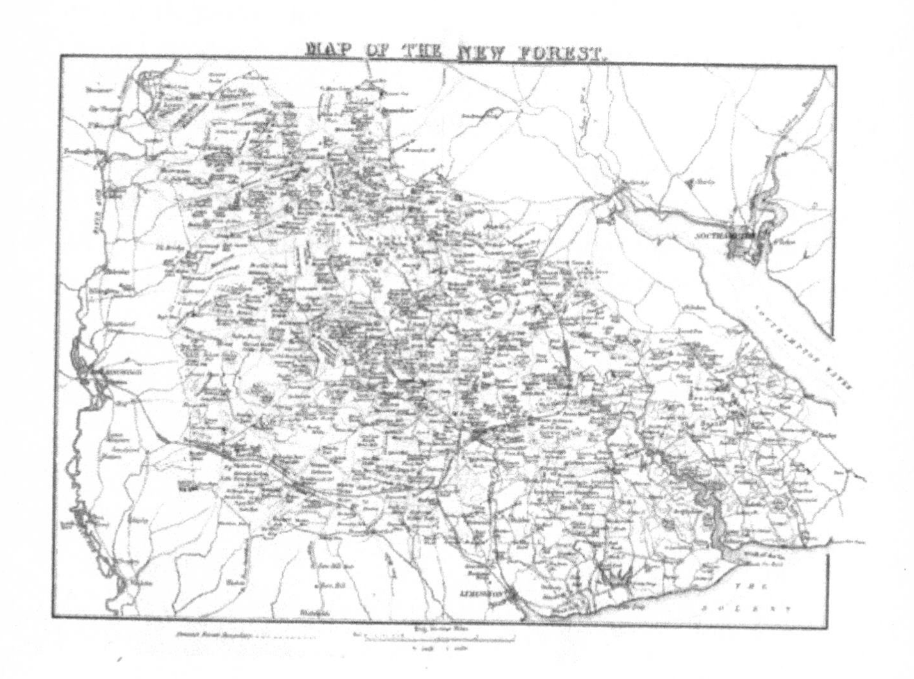

Footnotes

1. The Forest would afford a good field for deciding the controversy as to whether our tame pigs are descended from the European Wild Boar. (See *Proceedings of the Zoological Society*, 1861, p. 264; and *Annals and Magazine of Natural History*, Third Series, vol. ix. p. 415.) Certain it is that here are some breeds distinct in their markings. I must not, too, forget to mention *Coronella lævis* (Boie), which is found in the Forest, as also in Dorsetshire and Kent. This is the *Coronella austriaca* of Laurenti, and afterwards the *Coluber lævis* of Lacépede. It might be mistaken for the common viper (*Pelias berus*), but differs in not being venomous, as also from the ringed snake (*Natrix torquata*) in having a fang at the hinder extremity of its jaws, the peculiarity

of the genus *Coronella*. It feeds on lizards, which its fang enables it to hold; drinks a great deal of water; and Dr. Glinther, of the British Museum, to whom I am indebted for the above information, tells me that it crawls up the furze and low bushes to lick the rain off the leaves. For a list of the Lepidoptera of the New Forest see Appendix IV.
2. Vol. i p. 26
3. *Illustrations of the Eggs of British Birds*, by W. C. Hewitson, vol. i. p. 27.
4. As so few opportunities occur of weighing the eggs of the honey-buzzard and hobby, the following notes, most carefully made by Mr. Rake and myself, may not be without interest:—
Honey-buzzard's nest, taken June 16th, in a low fork of an oak-tree in Anses Wood, contained two fresh-laid eggs:—

First egg (apothecaries' weight)	1oz. 3dr. 1sc. 5gr.
Second egg (very slightly dinted)	1oz. 2dr. 2sc. 10gr.

Honey-buzzard's nest, taken June 24th, in Ravensnest Wood, near Brook, in the higher branches of a tall beech, overhanging the road. This nest had been deserted, and the two eggs were very much addled and hard set:—

First egg	1oz. 4dr. 0sc. 10gr.
Second egg	1oz. 3dr. 2sc. 10gr.

Hobby's nest, placed in a nest which, in 1861, had been occupied by a honey-buzzard, was taken in Prior's Acre, June 21st, and contained three fresh-laid eggs, now in Mr. Rake's cabinet:—

First egg	6dr. 0sc. 0gr.
Second egg	5dr. 2sc. 10gr.
Third egg (very slightly dinted)	5dr. 2sc. 0gr.

Hobby's nest, taken in South Bentley Wood, July 12, contained two eggs hard sat upon and addled:—

First egg	5dr. 2sc. 15gr.
Second egg (cracked)	5dr. 0sc. 14gr.

With these weights may be compared the following:—Egg, supposed to be that of a merlin, taken with two others which were broken, June 17th, 1862, near Alum Green, in the hole of a beech, rather sat upon, weighed 4dr. 1sc. 10gr. Two fresh-laid eggs of kestrels, taken at the same time, weighed 4d. 2sc. 15gr. Other eggs of kestrels, however, have weighed considerably more; and two others, also laid about the same time, came to 5dr. 5gr.

5. As the instances of the breeding of the merlin, especially under these circumstances, will always be very rare, I may as well add my own personal observations. In the spring of 1861 I received three eggs taken not far from the Knyghtwood Oak, and said to have been found in the hole of a beech. As I am not in the habit of paying any attention to the mere stories which are so plentiful, I did not, therefore, examine them with any attention, and put them aside as merely kestrel's. After, however, Mr. Farren's communication to me, I looked out particularly for this little hawk, but only once saw it in the open ground, near Warwickslade Cutting, from whence it flew up, perching for a moment on a

holly, and then making off to the woods. On June 4th, however, I observed a hen bird fly out of a hole, about twenty feet from the ground, in an old beech in Woolstone's Hill, on the east side of Haliday's Hill Enclosure. There were, however, no eggs. On the 5th I went again, and the bird, when I was about fifty yards from the tree, again flew off. Still, there were no eggs. I did not return till the 9th, when the nest, now pulled out of the hole, had been robbed. It was made of small sticks, and a considerable quantity of feather-moss, and some fine grass, and in general character resembled the nests of the bird found by Mr. Hewitson in Norway. In the holes were the bones of young rabbits, but these had, from their bleached appearance, been brought by a brown owl, who had reared her brood there in the previous summer. I afterwards learnt where the three eggs had been taken in 1861; but there was nothing, with the exception of a few sticks, in the hole, which was in this case about ten feet from the ground, and placed also in a beech on the edge of Barrowsmoor. Great caution, however, must be exercised regarding the merlin's eggs; for I am inclined to think that the kestrel, contrary to its usual practice, sometimes also breeds in the Forest in the holes of trees. The egg mentioned at p. 264, foot-note, brought to me on June 17th, 1862, I have every reason to believe is a merlin's, but could not quite satisfy myself as to the evidence.

6. For some account of the little owl (*Strix passerina*), see Appendix III. under the section of Stragglers, p. 314.

7. Vol. ii. p. 57.

8. Yarrell, vol. ii. p. 139.

9. Passed in the twenty-fourth year of Henry VIII., 1532. *Statutes of the Realm*, vol. iii., p. 425, 426. It should, however, be remembered that under the term chough was in former times included the whole of the Corvidæ. Shakspeare's "russet-pated choughs" are evidently jackdaws.

10. In Appendix III. is given a list of all the birds hitherto

observed in the New Forest District, as also more special information, which I thought would not interest the general reader.

APPENDIX I.

A GLOSSARY OF SOME OF THE PROVINCIALISMS USED IN THE NEW FOREST.

I COULD easily have expanded the following glossary to three times its size, but my object is to give only some specimens of those words which have not yet found their way into, or have not been fully explained in Mr. Halliwell's or Mr. Wright's dictionaries of provincialisms. The following collection is, I believe, the first ever made of the New Forest, or even, with the exception of the scanty list in Warner,[1] of Hampshire provincialisms, which of course to a certain extent it represents,—more especially those of the western part of the county. A separate work, however, would be needed to give the whole collection, and the following examples must here suffice.

Of course I do not say that all these words are to be found only in the New Forest. Many of them will doubtless be elsewhere discovered, though they hitherto, as here, have escaped notice. The time, however, for assigning the limits of our various provincialisms and provincial dialects has not yet arrived.

The use of the personal pronoun "he," as, throughout the West of England, applied to things alike animate and inanimate, and the substitution of "thee" for you, when the speaker is angry, or wishes to be emphatic, may be here noticed. In the Forest, too, as in parts of Berkshire, a woman when employed upon out-door work is sometimes spoken of in the masculine gender, as the Hungarians are falsely said to have done of their queen on a certain memorable occasion. The confusion of cases which has been noticed by philologists is here, as in other parts of England, rather the result of ignorance than a peculiar character of the dialect.

Adder's-fern. The common polypody (*Polypodium vulgare*), so called from its rows of bright spores. The hard-fern (*Blechnum boreale*) is known as the "snake-fern."

ALLOW, To. To think, suppose, consider. This word exactly corresponds to the American "guess" (which, by the way, is no Americanism, but used by Wiclif in his Bible: see Luke, ch. vii. v. 43), and is employed as often and as indefinitely in the New Forest. If you ask a peasant how far it is to any place, his answer nearly invariably is, "I allow it to be so far." "Suppose," in Sussex, is used much the same way.

Bell-heath. *See* Red-heath.

Bed-furze. The dwarf furze (*Ulex nanus*), which is very common throughout the Forest.

Black-heath. *See* Red-heath.

Black-heart, The. The bilberry (*Vaccinium Myrtillus*), the "whimberry" of the northern counties, which grows very plentifully throughout the Forest. It is so called, by a singular corruption, the original word being hartberry, the Old-English *heorot-berg*, to which the qualifying adjective has been added, whilst the terminal substantive has been lost, and the first totally misapprehended. To go "hearting" is a very common phrase. (See *Proceedings of the Philological Society*, vol. iii. pp. 154, 155.)

BRIZE. To press. "Brize it down," means, press it down. Is this only another form of the old word prize, preese, to press, crowd?

Boughy. A tree, which instead of running up straight is full of boughs, is said to be "boughy." It is also used generally of thick woods. Akin to it is the old word buhsomenesse, boughsomeness, written, as Mr. Wedgewood notices (*Dictionary of English Etymology*, p. 285), buxomeness by Chaucer.

Bower-stone, A. A boundary-stone. Called a "mere-stone" in some of the Midland Counties. Perhaps from the Keltic *bwr*, an inclosure, intrenchment; just as manor is said to be from *maenawr*, a district with a stone bound.

Bound-oak. *See* Oak, Mark-.

Brownies, The. The bees. *See* chap, xvi., p. 185.

Brow. Mr. Halliwell and Mr. Wright give this as a Wiltshire word, in the sense of brittle. In the New Forest it is applied only to short, snappy, splintering timber of bad quality.

Buck, The. The stag-beetle, so called from its strong horn-like *antennæ*. The children, when catching it, sing this snatch—

"High buck,
Low buck,
Buck, come down."

It is also called pinch-buck. The female is known as the doc. *See* "Bryanston Buck," in Mr. Barnes's *Glossary of the Dorsetshire Dialect*, appended to his *Poems of Rural Life*.

Bunch, A. A blow, or the effects of a blow; and then a blotch, burn, scald, pimple, in which latter senses "bladder" is also often used. The verb "to bunch," to strike, is sometimes heard. *See* Mr. Wedgewood (as before, p. 269) on its allied forms.

Cammock, The. (From the Old-English *cammec, cammoc, cammuc*.) The various species of St. John's-wort, so plentiful in the neighbourhood of the New Forest; then, any yellow flower, as the fleabane (*Pulica dysenterica*) and ragwort (*Senecio Jacobæa*). In Dorsetshire, according to Mr. Barnes, it only means the rest-harrow (*Ononis arvensis*).

Cass, A. A spar used in thatching, called in the Midland and North-Western Counties a "buckler." Before it is made into a cass, it is called a "spargad."

CATTAN, A. A sort of noose or hinge, which unites the "hand-stick" to the flail. It is made in two parts. The joint which joins the "hand-stick" is formed of ash or elm, whilst that which fits the flail is made of leather, as it is required to be more flexible near the part which strikes the floor. Mr. Wright and Mr. Halliwell give as a North-country word the verb "catton." to beat, with which there is evidently some connection.

CHILDAG, A. A chilblain. Often called simply a "dag," and "chilbladder."

CLEET, A. More generally used in the plural, as "cleets." Iron tips on a shoe. Hence we have the expression, "to cleet oxen," that is, to shoe them when they work.

CLOSE. Hard, sharp. "It hits close," means it hits hard.

COTHE. (From the Old-English "coða, coðe.") A "cothe sheep," means a sheep diseased in its liver. The springs in the New Forest are said "to cothe" the sheep—that is, to disease their livers. Hence we have such places as "Cothy Mead," and "Cothy Copse." Mr. Barnes (as before) gives the form "acothed," as used in Dorsetshire.

Crink-crank. "Crink-crank words" are long words—*verba sesquipedalia*—not properly understood. (See *Proceedings of Philological Society*, vol. v. pp. 143-148.)

Crow-peck, The. The Shepherd's needle (*Scandix-pecten Veneris*); called also "old woman's needle." There is a common saying in the New Forest, that "Two crow-pecks are as good as an oat for a horse;" to which the reply is, "That a crow-peck and a barley-corn may be."

Crutch, A. (From the Friesic *krock*, connected with the Old-English *crocca*, our crock). A dish, or earthenware pipkin. We daily in the New Forest and the neighbourhood hear of lard and butter crutches. The word "shard," too, by the way, is still used in the Forest for a cup, and housewives still speak of a "shard of tea."

Cuttran, A. A wren; more commonly called a "cutty;" which last word Mr. Barnes gives in his *Glossary of the Dorsetshire Dialect*, p. 331, but which is common throughout the West of England. As Mr. Barnes, p. 354, observes, the word is nothing more than cutty wren—the little wren. (*See* "Kittywitch," *Transactions of Philological Society*, 1855, p. 33.)

Decker, or Dicker, To. One of the old forms of to deck; literally, to cover; from the Old-English "þeccan;" in German, *decken*. It now, however, only signifies to ornament or spangle. A lady's fingers are said to be deckered with rings, or the sky with stars.

Deer's-milk. Wood-spurge (*Euphorbia amygdaloides*). So called from the white viscous juice which exudes from its stalks when gathered.

Dount, To. To dint, or imprint. Formed, as Mr. Wedgewood remarks, of the kindred words, dint, dent, dunt, by an onomatopoëtic process. We find the word in an old song still sung in the New Forest, "A Time to remember the Poor:"—

"Here's the poor harmless hare from the woods that is tracked,
And her footsteps deep dounted in snow."

Dray, A. A prison; "the cage" of the Midland districts. Curiously enough the old poet William Browne, as also Wither, speaks of a squirrel's nest as a "dray"—still used, by-the-by, in some counties—which in the New Forest is always called a "cage." In this last sense Mr. Lower adds it to the glossary of Sussex provincialisms (*Sussex Archæological Collections*, vol. xiii., p. 215). I may further note that at Christmas in the Forest, as in other wooded parts of England, squirrel-feasts are held. Two parties of boys and young men go into the woods armed with "scales" and "snogs" (see chap. xvi. p. 182). to see who will kill the most squirrels. Sometimes as many as a hundred or more are brought home, when they are baked in a pie. Their fur, too, is sought after for its glossiness.

DRUM, IVY-, An. The stem of an ivy tree or bush, which grows round the bole of another tree.

Drunch, To. To draw up, press, squeeze. We find the substantive "drunge," with which it is evidently connected, given in Wright as a Wiltshire pronunciation for pressure, or crowd. Mr. Barnes also, in his *Glossary of the Dorsetshire Dialect*, p. 235. gives the forms "dringe" or "drunge," to squeeze or push.

ELAM, An. An handful of thatch. Common both in the New Forest and Wiltshire. In the former three elams make a bundle, and twenty bundles one score, and four scores a ton. In the latter the measurement is somewhat different, five elams forming a bundle.

Fessey. (From the Old-English *fus*, ready, prompt, quick). Proud, upstart. In the glossaries of Wright and Halliwell we find "fess" given as the commoner form.

FETCH, To. Used with reference to churning butter. "To fetch the butter," means, to raise the cream into a certain consistency.

Fire-bladder. A pimple, or eruption on the face. *See* "bunch."

FLISKY. Small, minute. Used especially of misty rain.

Flitch, or quite as often Fritch. (From the Old-English *flit*, or *geflit*). Not only as explained in the glossary of Wiltshire, impertinent, busy, but, by some *boustrophêdon* process, good-humoured. "You are very flitch to-day," that is, good-natured.

Fluders. Worms, which on certain land get into the livers of sheep, when the animal is said to be "cothed." Called also "flukes," and "flounders." *See* the word "cothe."

GAIT, A. A crotchet, or, as the vulgar expression is, a maggot. Used always in a deprecatory sense. When a person has done anything foolish he says, "this is a gait I have got."

GETTET. Sprung, or slightly cracked. Used throughout the West of England.

GIGGLE, To. To stand awry or crooked. Said especially of small things, which do not stand upright.

Glutch, To. (From the French en-gloutir). Not simply, to swallow or gulp, as explained in the glossaries, but more especially to keep down or stifle a sob.

Gold-heath, The. The bog-moss (*Sphagnum squarrosum*), which is used in the New Forest to make fine brooms and brushes.

Gold-withey, The. The bog-myrtle, or English mock-myrtle (*Myrica Gale*), mentioned in Mr. Kingsley's New Forest ballad,—

> "They wrestled up, they wrestled down,
> They wrestled still and sore;
> Beneath their feet, the myrtle sweet,
> Was stamped in mud and gore."

It grows in all the wet places in the Forest, and is excessively sweet, the fruit being furnished with resinous glands. It is said to be extensively used in drugging the beer in the district.

Graff, or grampher. *See* Wosset.

Gross. Often used in a good sense for luxuriant, and applied to the young green crops, just as "proud," and "rank," or rather "ronk," as it is pronounced, are in the Midland Counties.

Gunney. To look "gunney" means, to look archly or cunning. There is also the verb "to gunney." "He gunneyed at me," signifies, he looked straight at me.

Hacker, Furze-, The. The whinchat, so called from its note, which it utters on the sprays of the furze.

Hame. There is a curious phrase, "all to hame," signifying, broken to pieces, used both here and in Wiltshire. Thus the glass, when broken, is said to be "all to hame," that is, "all to bits." The metaphor has been taken from "spindly" wheat on bad ground running to halm, from the Old-English *healm*, now the West-Saxon peasant's "hame." "All to," I may add, is used adverbially in its old sense of entirely, quite, as we find it in Judges ix. 53.

HARL, The. The hock of a sheep.

Harvest-lice. The seeds of the common agrimony (*Agrimonia eupatoria*) and heriff (*Cheniopodium album*). *See* Clivers, chap. xv. p. 166.

Hell. A dark place in the woods. *See* chap. x. p. 110.

Herder. A sieve. *See* chap. xvi. p. 185, foot-note.

Hill-trot, The. The wild carrot (*Daucus carota*), used also in Wiltshire. Most probably a corruption of eltrot, old rot, oldroot, and so from the Old-English. These last forms are given in Mr. Barnes' *Glossary of the Dorset dialect*, p. 336.

Hoar-withey. The whitebeam (*Sorbus Aria*), which, with its white leaves, is very conspicuous in the Forest. We find the word used in the perambulation of the Forest in the twenty-second year of Charles I.,—"by the road called Holloway, and from thence to Horewithey, in the place whereof (decayed) a post standed in the ground." It is exactly the same as the "har wiðig" of the Old-English. It is called also, but more rarely, the "white rice." *See* chap. xvi. p. 183.

Hoo, To. To simmer, boil; evidently formed, like so many other words, by an onomatopöetic process (*See* chap. xvi. p. 186). There is also the phrase, "the kettle is on the hoo," that is, to use a vulgarism, on the simmer, or boil.

HOOP, To go a. To go where you like. "He is going a hoop," means, he is going to the bad.

Hum-water. A cordial which is made from the common horse-mint (*Mentha aquatica*). Does "hum" here mean strong, as it is used in some counties with reference to beer? *See* chap. xv. p. 166.

JOSEPH'S WALKING-STICK. The Joseph's-ladder of the Midland Counties, common in all the cottage gardens round the Forest. It is curious to notice, amongst our peasantry, the religious element in the names of both the wild and cultivated flowers derived from Catholic times. Thus we have ladies' cushions, and ladies' tresses, and St. Peter's-wort, and St. John's wort, besides the more common plants, such as marygolds and ladysmocks, which every one can remember.

KITTERING. Weak. The more North-country word "tuly" is also heard in the same sense.

LANCE, To. To jump, leap, or bound. Used especially of the Forest deer, which in dry weather are said "to lance" over the turf.

LARK'S LEES, or LEASE, A. A piece of poor land fit only for larks, or, as the peasantry of the Midland Counties would say, only "fit to bear peewits." Mr. Halliwell gives the form "lark leers," as a Somersetshire phrase; but the above expression may be daily heard in the New Forest.

LOUSTER. Noise, disturbance. "What a louster you are making," signifies, what a confusion you are causing.

Lug-stick. *See* Rug-stick.

Mallace, The The common mallow (*Malvus sylvestris*). Formed like bullace, and other similar words.

Margon. Corn chamomile (*Anthemis arvensis*). Culled "mathan," throughout the Anglian districts.

Mark-oak, *See* Oak.

Mokin, or more generally in the plural, Mokins. Coarse gaiters for defending the legs from the furze. *See* chap. xv. p. 162.

Muddle, To. To fondle, caress, to rear by the hand. Hence we get the expression "a mud lamb," that is, a lamb whose mother is dead, which has been brought up by hand, equivalent to the "tiddlin lamb" of the Wiltshire shepherds. *See* Wosset.

Oak, Mark-, A. The same as a "bound-oak," or boundary oak or ash, as the case may be, so called from the ancient cross, or mark, cut on the rind. As Kemble notices (*The Saxons in England*, vol. i., appendix A. p. 480), we find in Cod. Dipl. No. 393, "on ðán merkeden ók," to the marked oak, showing how old is the name. I have never met in the New Forest with an instance of a "crouch oak" (from *crois*), such as occurs at Addlestone in Surrey, and which is said to have been the "bound-oak of Windsor Forest (*See The Saxons in England*, as before, vol i. chap. ii. p. 53, fool-note). The "bound-oak," marked in the Ordnance Map near Dibden, has fallen, but we find the name preserved in the fine old wood of Mark Ash, near Lyndhurst. In the perambulation of the Forest in the 29th year of Edward I. we read of the Merkingstak of Scanperisgh. The various eagle-oaks in the Forest are comparatively modern, and must not be confounded with the eagle-oak mentioned by Kemble (as above, vol. i. p. 480).

Omary Cheese. An inferior sort of cheese, made of skim-milk, called in most parts of England "skim Dick." *See*, further on, the word Rammel, and also Vinney, chap. xvi. p. 190.

Once. Sometime. "I will pay you once this week," does not mean in contradistinction to twice, but I will pay you sometime during the week.

Overrunner, An. A shrew mouse, which is supposed to portend ill-luck if it runs over a person's foot. In Dorsetshire it is called a "shrocop," where the same superstition is believed. *See* Barnes' *Glossary of the Dorset Dialect*, p. 382.

Panshard, or Ponshard, A. Rage, anger. "You have no need to get in a panshard," is a most common saying. *See* "peel," further on.

Patchy. Testy. Said of people who proverbially "blow hot and then cold."

Peel, A. A disturbance, noise. "To be in a peel," means, to be in a passion. Used in much the same sense as the word "pelt," which is rightly explained in the glossaries as anger, noise, rage, though it is, perhaps, more spoken of animals than "peel." "What a pelt the dog is making," that is, barking, would be said rather than "peel."

Picked. Sharp, pointed. "A picked piece," means a field with one or more sharp angular corners.

Pity. Love. "Pity is akin to love," says Shakspeare, but in the West of England it is often the same.

Plash, A. A mill-head. It is often used conjointly with another word, as Winkton Plash.

Puck, To. To put up sheaves, especially of barley and oats, which are called "pucks." Used throughout the West of England in contradistinction to "hiling," applied only to wheat, which is placed in "hiles." In Dorsetshire, however, this last operation is called "stitching." *See* the word "stitch" in Mr. Barnes' *Glossary of the Dorsetshire Dialect*, p. 391.

Quar, A. The udder of a cow or sheep, when hard after calving or lambing. Beer also is said to be "quarred," when it drinks hard or rough.

Quat-vessel, The. The meadow-thistle (*Cardans pratensis*), which is common in the New Forest.

Rammel Cheese. The best sort of cheese, made of cream and new milk, in contradistinction to Omary cheese, which see.

Rammucky. Dissolute, wanton. "A rammucky man," means a depraved character.

Ramward, or rather, ramhard. To the right. A corruption of framward, or fromward. So "toard," or "toward," means to the left, that is, towards you. Both words are used throughout the West of England, and are good examples of what Professor Müller would call "phonetic decay." With them may be compared the sailor's terms "starboard" (*questo bordo*, this side), and larboard (*quello bordo*, that side). *See*, too, Miss Gurney on the word "woash," which in the Eastern Counties is equivalent to "ramward." *Glossary of Norfolk Words*: *Transactions of the Philological Society*, 1855, p. 38.

Rantipole, The. The wild carrot (*Daucus carota*), so called from its bunch of leaves. Used also in Wiltshire. *See* Hill-trot.

Red Heath. The three heaths which grow in the New Forest—*Erica tetralix*, *Erica cinerea*, and *Calluna vulgaris*,— are respectively known as the bell, black, and red heaths.

Reiaves. The boards or rails put round waggons, so as to enable them to take a greater load. Used throughout the West of England. *See* Mr. Barnes' Glossary under the word Riaves, p. 375.

Rick-rack. This is only used of the weather, as "rick-rack weather," that is, stormy, boisterous weather, and far stronger in meaning than the more common phrase, "cazalty weather." It is evidently from the Old-English *ree*, vapoury, cloudy weather, and well serves to explain the meaning of Shakspeare's "rack," a cloud, in the well-known passage in the *Tempest* (Act iv. sc. 1), which has given rise to so much controversy. Miss Gurney (*Transactions of the Philological Society*, 1855, p. 35), notices that "rack" is used in Norfolk for mist driven by the wind.

RONGE, To. To kick, or play, said of horses.

RUBBLE, To. To remove the gravel, which is deposited throughout the Forest in a thick layer over the beds of clay or marl. The gravel itself is called "the rubblin."

Rue. A row, or hedgerow. *See* chap. v. p. 56. In the Forest some of the embankments, near which perhaps the Kelts and West-Saxons lived, are called Rew- and Row-ditch. I have, too, heard of attics being called "lanes," possibly having reference to the "ruelle" by which the space between the curtains was formerly called.

RUG-STICK, also called a Lug-stick. A bar in the chimney, on which "the cotterel," or "iron scale," or "crane," as it is also called, to which the kettle or pot is fastened, hangs. We find the word still used in America as the "ridgepole" of the house, which helps us at once to the derivation.

Scale, or squoyle. *See* chap. xvi. p. 182.

Scull, A. (From the Old-English *scylan*, and so, literally, a division). A drove, or herd, or pack of low people, always used in an opprobrious sense. It is properly applied to fish, especially the grey mullet which visits the coast in the autumn, and so metaphorically to beggars who go in companies. Milton uses the word

> "————sculls that oft
> Bank the mid sea."
> *Paradise Lost*, Book vii.

Shakspeare, too, speaks of "scaled sculls" (*Troilus and Cressida*, Act v. sc. 5). The expression "school of whales," which we so often find in Arctic and whaling voyages is nothing but this word slightly altered. According to Miss Gurney's Glossary of Norfolk words (*Transactions of the Philological Society*, 1855), the word "school" is applied to herrings on the south-eastern coast. Juliana Berners, in the *Book of St. Albans*, curiously enough says that we should speak of "a sculke of foxes, and a sculle of frerys."—Quoted in Müller's *Science of Language*, p. 61.

SETTY. Eggs are said to be "setty" when they are sat upon.

SHAMMOCK, To. To slouch. "A shammocking man" means an idle, good-for-nothing person. Applied also to animals. "A shammocking dog," means almost a thievish, stealing dog, thus showing how the word is akin to shamble, scamble, which last verb also signifies to obtain any thing by false means.

SHEAR, AFTER-. The. The second crop of grass. Called in the Midland Counties "the eddish," and also the "latter-math," or "after-math."

Sheets'-axe, A. An oak apple. *See* chap. xvi. p. 183.

Shelf, A. A bank of sand or pebbles, or shallow in a river, or even the ford itself. Milton uses the word in *Comus*:—

"On the tawny sands and shelves."Hence we got the adjective "shelvy," also in common use, and employed by Falstaffe—"The shore was shelvy and shallow" (*The Merry Wives of Windsor*, Act iii., sc. 5). It is this latter word, which Mr. Halliwell and

Mr. Wright must mean instead of "shelly," and which they define as "an ait in a river." The word is probably from the same Scandinavian root as shoal.

SHIM. Lean. "He's a shim fellow," that is, thin. It is used, I see from Mr. Cooper's glossary, for a shadow, in the western division of Sussex; and I think I have somewhere met with it in the sense of a ghost.

Shoak, Shock, Shuck, Off, To. To break off short. Thus gravel is said to shock off at any particular stratum, or "list," or "scale," as it would be called. *See* the following word.

SHOCK, A. Not applied merely to corn, but to anything else. "A shock of sand" means a line or band of sand, called also a "list," or "lissen," or "bond," or "scale," and sometimes "drive:" which last, however, has a more particular reference to the direction of the stratum.

SIZE. Thickness, consistency. "The size of the gruel" means its consistency.

SKIMMER-CAKE, A. A small pudding made up from the remnants of another, and cooked upon a "skimmer," the dish with which the milk is skimmed. Nearly equivalent to the "girdle-cake," north of England.

SKROW. Shattered or battered.

SLAB, A. A thick slice, lump, used like squab, which see. Thus we hear of "a slab of bacon," meaning a large piece. Opposed to "snoule," which signifies a small bit.—"I have just had a snoule," means I have only had a morsel.

SLINK, A. "A slink of a thing," in which phrase the word is only found, is alike applied to objects animate or inanimate, and means either a poor, weak, starved creature, or anything which is small and not of good quality.

SLUT, A. A noise, sound. "A slut of thunder," means a chip or peal of thunder. It is in this sense that the word is most generally used.

Snake-fern. The hard-fern (*Blechnum boreale*). *See* "Adder's-Fern."

Sniggle, To. To snarl. *See* chap, xvi., p. 186. Sniggle, A. An eel peculiar to the Avon. *See* chap, xii., pp. 125, 126.

SPELL, A. A fit, or start. Pain is said to come and go by "spells," that is, by shocks at recurring intervals.

Spene, A. In its first sense, like the Old-English *spana*, an udder of a cow. In its second, the rail of a gate or stile.

Spine-oak. The heart of oak. This phrase points to the true derivation of "heart of oak." The common theory Mr. Wedgewood has rightly classed under the head of "False Etymologies." *See Transactions of the Philological Society*, 1855. No. 6, pp. 62, 63.

Spire-bed, A. A place where the "spires," that is, the reed-canary grass (*Phalaris arundinacea*), grow; exactly equivalent to the Old-English *hreodbedd*. On the outskirts of the New Forest at Redbridge, formerly Redford—Hreodford, literally, the ford of reeds—the Test is to this day full of the same "spires," from which our forefathers gave the place its name. The river Caundle, in Dorsetshire, still, too, full of spire-beds, tells of a similar derivation, not from the Teutonic, but the Keltic. The phrase "spire-bed," or "spear-bed field," is very common, meaning a particular field, near where the "spires" grow, which are used by plasterers and thatchers in their work.

SPITH. (Another form of pith, from the Old-English "piða"). Strength, force.

SPRACK. Not only quick, lively, brisk, active, as given in the glossaries, but neat, tidy. Used also in this last sense in Wiltshire.

Spratter. The common guillemot (*Uria trone*). In Norfolk (*see Transactions of the Philological Society*, 1855, p. 37) we have "sprat-mowe," for a herring-gull; and in Kent, "sprat-loon," for one of the grebes.

SQUAB, A. Anything large. Thus "a squab of a piece," is constantly used in this sense. In a different meaning it is confounded with squat. So a thickset, heavy person is called a "squab."

Squoyles. Glances. *See* chap, xvi., p. 182.

Stabble. Marks, footprints, always used in the plural. This is another of those onomatopoëtic words which Mr. Wedgewood might add to the forms step, stamp, stipple, all derived by a similar process. (*See* the Introduction to his *Dictionary of Etymology*, p. x.) In an old rhyme, common in the New Forest, upon a hailstorm, we find the word:—

"Go round the ricks,
And round the ricks,
And make as many stabble

As nine score sheep."

STARKY. Used particularly of land which is stiff or unworkable, especially after rain, and opposed to "stoachy," which signifies muddy, as in the common expression, "What a dreadful stoachy piece of ground."

THRIFTY. Still used in its old derivative sense of thriving, and so flourishing. Once or twice I have heard it applied to physical health, in the sense of being well, or "pure," as is the more common saying.

Tine, To. To tine a candle, does not now so much mean to light, from the Old-English *tendan*, to set on fire, as to snuff it.

TUFFET, A. A lump of earth, or hillock. Hence we have "tuffety," in the sense of uneven, or covered with hillocks.

Tuly. Weak, ailing. More common in the north of England. *See* "Kittering."

TWIDDLE, To. To whistle. "The robins are twiddling," is a common phrase, and which fact is said to be a sign of rain.

Vinney-Cheese. *See* chap, xvi., p. 190.

Wag, A. A breath, a slight wind. "A wag of air," means a gentle draught of air. In Dorsetshire we still have "wag-wanton" applied to the quaking-grass (*Briza media*). *See* Barnes' *Glossary of the Dorsetshire Dialect*, p. 404.

Wase, A. A very small bundle of straw, more particularly a wisp for cleaning a horse. Used also, according to Mr. Cooper, in Sussex.

Water-tables. The side dikes along the road, which carry off the water. Common throughout the West of England.

Weald, To. To bring corn or hay into swathe, before putting it, as it is called, into "puck," which see.

Wean-gate, A. (From the Old-English wæn-geat, literally, the waggon-door.) The tail-board, or ladder of a waggon.

Well-crook, A. A stick for ladling the water out of the shallow Forest pools and wells. Called in the Midland and Northern Counties a "lade-gorn."

Wimble, A. In addition to auger, as given in Wright and Halliwell's dictionaries, an instrument with which to take up faggots or trusses of hay.

Wine, Cider-. Often used instead of cider.

Wivvery. Giddy. "My head is wivvery," is no uncommon expression. To wivver, given by Wright and Halliwell as used in Kent, is more especially employed here of the quivering flight of hawks, particularly of the kestrel and hen-harrier.

Wosset, A. A small ill-favoured pig. The smallest pig in a "trip," to use a West-Country term for a litter, is known as the "doll," the same as the "nessle-tripe" of Dorsetshire; whilst a pig brought up by hand is called a "graft"," or "grampher," equivalent to "mud," in the phrase "mud-lamb," or "mud-calf," as also "sock," and "sockling," and "tiddling," used in various counties.

Yape, To. Not merely to gossip, as given by Mr. Cooper in his *Sussex Glossary*, but to loiter. To yape about is used very much as is shammock which see.

YAW, To. To chop, reap. Used of cutting corn, peas, or beans. "Hacking," however, is generally the term applied to harvesting the last, when the reapers use two hooks, one to cut, and the other, an old one, to pull up the halm.

The Staple Cross, near Christchurch.

Footnotes

1. Collections for the *History of Hampshire*, by Richard Warner, vol. iii., pp. 37, 38. A brief list of Hampshire words will also be found in *Notes and Queries*, First Series, vol. x., No. 250, p. 120. Mr. Halliwell, in his account of the English Provincial Dialects, p. xx., prefixed to his *Dictionary of Archaic and Provincial Words*, mentions a MS. glossary of the provincialisms of the Isle of Wight, by Captain Henry Smith of which he has made use.

APPENDIX II.

THE FLOWERING PLANTS OF THE NEW FOREST DISTRICT.

These lists are not by any means put forward as exhaustive. Subsequent investigations must very much increase them. Still, I trust they will be found sufficient for botanists to generalize from, and useful as guides to beginners. To the kindness of the Rev. H. M. Wilkinson, of Bisterne, I am much indebted, as will be seen, for many new species and localities, as also for the special arrangement of the *Gramineæ*, *Cyperaceæ*, and *Juncaceæ*.

The nature of the country will best help us to make the divisions. First, we have the true Forest district, with its heath, and bog, and woodland plants; and next the valley of the Avon, with its meadow-flowers; and, thirdly, the littoral plants, which we will at once take.

GLAUCIUM LUTEUM, Scop., Yellow-horned Poppy. Leap. Eaglehurst, 46.[1]

CAKILE MARITIMA, Scop., Purple Sea-rocket. The sea-shore, Mudeford, 55.

Crambe maritima, Lin., Sea Kale. The sea-shore near Calshot and Eaglehurst, where, as Bromfield remarks (*Flora Vectensis*, p. 48), the young shoots are bleached by being covered with shingle, and then sent to the Southampton market, 56.

COCHLEARIA OFFICINALIS, Lin., Common Scurvy Grass. Hurst Castle, 72.

COCHLEARIA ANGLICA, Lin., English Scurvy Grass, Mudeford.

R. Stevens, Esq., 72 d.

RAPHANUS MARITIMUS, Sm., Sea Radish. Mudeford, 124.

SILENE MARITIMA, With., Sea Bladder Campion. The Shingles. Hurst Castle, 153.

HONCKENEJA PEPLOIDES, Ehrh., Sand Chickweed. Common on the coast, 173.

SPERGULARIA MARINA, Camb., Sea Spurrey. Mudeford, 174.

ALTHÆA OFFICINALIS, Lin., Marsh Mallow. Salt marshes of the Beaulieu river, 208.

LAVATERA ARBOREA, Lin., Tree Mallow. Hurst Castle, where Ray saw it. See, however, Bromfield in Phytologist, vol. iii. p. 270; 210.

ANTHYLLIS VULNERARIA, Lin., Common Lady's Fingers. Barton Cliffs, 257.

Tamarix gallica, Sm. "On the beach near Hurst Castle." Garnier and Poulter. Milford. Probably naturalized, as on the opposite coast near Yarmouth. "The Lymington Salterns," Rev. H. M. Wilkinson. *See*, however, Bromfield, in *Phytologist*, vol. iii. p. 212; 392.

ERYNGIUM MARITIMUM, Lin., Sea Holly. Mudeford, 444.

FŒNICULUM VULGARE, Gærtn, Common Fennel. Purewell Road, Christchurch, 476.

APIUM GRAVEOLENS, Lin., Wild Celery. "Marchwood," W. A. Broomfield. "Mudeford and Beaulieu," Rev. H. M. Wilkinson, 450.

ŒNANTHE LACHENALII, Gmel., Lachenal's Dropwort. "Mudeford," Rev. H. M. Wilkinson, 471*

CARDUUS TENUIFLORUS, Curt., Small-flowered Thistle. Lanes near the sea-coast, 597.

Artemisia maritima, Lin., Sea Wormwood. "The coast." W. Pamplin. "Salt marshes near Millbrook," W. A. Bromfield; quoted in the *New Botanist's Guide*, 624.

ASTER TRIPOLIUM, Lin. Sea Starwort. Very common in the

rivers at Beaulieu and Lymington, 641.

INULA CRITHMOIDES, Lin., Golden Samphire. Key Haven and Hurst Beach, where Ray saw it, 657.

CONVOLVULUS SOLDANELLA,Lin., Sea Bindweed. Hurst Castle. Mudeford, 731.

GLAUX MARITIMA, Lin., Sea Milkwort or Glasswort. Hurst Castle, Beaulieu Estuary, 894.

ARMERIA MARITIMA, Aut., Common Thrift. Hordle and Barton Cliffs, Beaulieu Estuary, 895.

Statice Limonium, Lin., Sea Lavender. On this and S. rariflora, see Bromfield, in *Phytologist*, vol. iii. p. 742; 897.

PLANTAGO MARITIMA, Lin., Sea Plantain. The Beaulieu Estuary, 904.

CHENOPODIUM OLIDUM, Curt., Stinking Goosefoot. Mr. Wilkinson gives "the seaside, Beaulieu," 908.

ATRIPLEX PORTULACOIDES, Lin. Hurst Castle, where I first saw it in 1859, with Mr. Lees, 918.

ATRIPLEX BABINGTONII, Wds., "Mudeford," Rev. H. M Wilkinson, 921.

ATRIPLEX LITTORALIS, Lin., Grass-leaved Sea Orache. Estuary of the Beaulieu river, 924.

BETA MARITIMA, Lin., Sea Beet. Mudeford, 925.

SALSOLA KALI, Lin., Prickly Saltwort. The seashore, Mudeford, 926.

SCHOBERIA MARITIMA, Mey., Sea Goosefoot. Estuary, 927.

SALICORNIA HERBACEA, Lin., Jointed Glasswort. "The Beaulieu river," Rev. H. M. Wilkinson, 939.

Polygonum maritimum, Lin., Sea Knot Grass. "Mudeford," Borrer, C. C. Babington. (*See* Watson's *New Botanist's Guide*, Supplement, vol. ii. p. 570.) The Rev. W. M. Wilkinson has found it on the other side of the harbour at Hengistbury Head, 940.

POLYGONUM RAII, Bab. "Mudeford," Borrer, and R. Stevens, Esq., 940*

ASPARAGUS OFFICINALIS, Lin., Common Asparagus. "At Christchurch," Garnier and Ponlter.

TRIGLOCHIN MARITIMUM, Lin., Sea Arrow Grass. Marshes of the Beaulieu river, 1115.

ZOSTERA MARINA, Lin., Narrow Grass Wrack. Southampton water, Hythe, 1137.

JUNCUS MARITIMUS, Sm., Lesser Sharp Sea Rush. Beaulieu river, 1154.

Scirpus savii, S. and M., Savis' Club Rush. *See* Bromfield, in *Phytologist*, vol. iii. p. 1030; 1187.

SCIRPUS MARITIMUS, Lin., Salt Marsh Club Rush. Mudeford, 1190.

CAREX EXTENSA, Good., Long Braeteated Carex. "The Beaulieu river," Rev. H. M. Wilkinson, 1235.

Ammophila arundinacea, Host., Sea Reed. The loose sand, Mudeford, where it grows with *Triticum junceum*, 1293.

GLYCERIA MARITIMA, M. and K., Sea Hard Grass. Mudeford, 1323.

GLYCERIA LOLIACEA, Watson, Dwarf Sea-wheat Grass. "Mudeford. On the New Forest side of the Avon, which is the only place I have ever seen it" Rev. H. M. Wilkinson, 1327.

TRITICUM JUNCEUM, Lin., Rushy Sea-wheat Grass. Mudeford, 1362.

Hordeum maritimum, With., Sea Barley. Very common along the whole of the east coast. "By the roadside from Cadenham" (more probably Hythe) "to Marchwood," W. A. Bromfield. *See* Watson's *New Botanist's Guide*, vol. ii., p. 571.; 1369.[2]

LEPTURUS FILIFORMIS, Trin., Sea Hard-grass. Mudeford, 1371.

In the next division are placed more especially those plants which either grow only in the Forest, or form a peculiar feature in its landscapes, such as *Eriophorum angustifolium*, *Gentiana Pneumonanthe*, *Drosera rotundifolia*, and *intermedia*, *Narthecium ossifragum*, *Melittis Melissophylium*, and the *Carices*,

Aivæ, and *Agrostes* generally. The rest will be found in the third division, as common both to the Forest and the adjoining districts. As the Ferns and St. John's-worts have been so fully mentioned in Chapter XXI., they will not be again noticed.

ANEMONE NEMOROSA, Lin., Wood Anemone, 6.

RANUNCULUS AQUATILIS, Lin., Water Crowfoot. Streams and pools, not of course confined to the Forest, but still a conspicuous feature, 11.

Ranunculus tripartitus, D. C, Three-parted-leaved Crowfoot, "with *Limosella aquatica*, in splashy places by the roadside, just beyond the bridge, as you leave Brockenhurst for Lyndhurst," H. C. Watson, in a private letter, 11.*

RANUNCULUS HIRSUTUS, Curt., Hairy Crowfoot. Roads in the Forest, 22.

CALTHA PALUSTRIS, Lin., Common Marsh Marigold. Forest pools; but, of course, in the district generally, 26.

Aquilegia vulgaris, Lin., Common Columbine. Very common round Wootton, but may be found with *Hypericum androsæmum* in the old woods of Mark Ash, Gibb's Hill, Winding Shoot, and Boldrewood, 31.

NYMPHÆA ALBA, Lin., White Water Lily. Forest streams. Not so common as the next, but still a feature, 36.

NUPHAR LUTEUM, Sm., Yellow Water Lily. In the Avon, and elsewhere in the district, 37.

VIOLA CANINA, Sm., Dog's Violet. The violet of the Forest, but, of course, common in the district, 135.

Viola lactea, Sm., Cream-coloured Violet. "Near Boldre," W. A. Bromfield. *See* Watson's *New Botanist's Guide*, vol. ii., p. 567; 135.*

DROSERA ROTUNDIFOLIA, Lin., Round-leaved Sundew. Everywhere in the Forest, 138.

Drosera intermedia, Hayn., Narrow-leaved Sundew. Though not so common as *rotundifolia*, it is equally distributed throughout the Forest district, 139.

POLYGALA VULGARIS, Lin., Common Milkwort, 141.

MŒNCHIA ERECTA, Sm., Upright Mœnchia. Common, 166.

SAGINA SUBULATA, Wimm., Ciliated Awl-shaped Spurrey, 170.*

SPERGULARIA RUBRA, St. Hilaire, Purple Sandwort, 175.

CERASTIUM SEMIDECANDRUM, Lin., Little Mouse-ear Chickweed, 194.

CERASTIUM TETRANDRUM, Curt., Four-cleft Mouse-ear Chickweed, 194.*

LINUM ANGUSTIFOLIUM, Huds., Narrow-leaved Flax, 201.

RADIOLA MILLEGRANA,, Sm., Thyme-leaved Flax-seed. Common. The Rev. P. Somerville pointed it out to me in Beacon Burney, growing close to the sea, 203.

TILIA INTERMEDIA, D. C, Common Lime, 212.

ACER CAMPESTRE, Lin., Field Maple. Rather plentiful in some of the woods, 225.

GERANIUM PRATENSE, Lin., Meadow Crane's-bill. On a rubbish heap, near Alum Green, where it had been naturalized, 231.

OXALIS ACETOSELLA, Lin., Wood-sorrel. Very common, 243.

EUONYMUS EUROPÆUS, Lin., Spindle Tree. Here and there a specimen may be seen, as at the north side of Wootton Enclosure, near the Osmanby Ford River, 245.

RHAMNUS FRANGULA, Lin., Alder Buckthorn, 247.

SPARTIUM SCOPARIUM, Lin., Common Broom, 248.

ULEX EUROPÆUS, Lin., Furze, 249.

ULEX NANUS, Forst., Dwarf Furze. If any one wishes to see the difference between this and Europaus he should visit the Forest at the end of August or the beginning of Sept., 250.

GENISTA TINCTORIA, Lin., Dyers' Green Weed. Common on the southern parts of the Forest, 251.

GENISTA ANGLICA, Lin., Petty Whin. Everywhere, 253.

TRIFOLIUM STRIATUM, Lin., Soft Knotted Trefoil, 277.

TRIFOLIUM FRAGIFERUM, Lin., Strawberry-headed Trefoil. Ashley Common, 280.

TRIFOLIUM GLOMERATUM, Lin., Smooth round-headed Trefoil, 278.

OROBUS TUBEROSUS, Lin., Common Bitter Vetch, 312.

PRUNUS SPINOSA, Lin., Sloe-tree, 314.

PRUNUS AVIUM, Lin., Wild Cherry. Burley, 316.*

POTENTILLA TORMENTILLA, Schk., Common Tormentil, 332.

COMARUM PALUSTRE, Lin., Purple Marsh Cinquefoil. Bog of the Osmanby Ford River, below Wootton Enclosure, 334.

FRAGARIA VESCA, Lin., Strawberry, 335.

RUBUS IDÆUS, Lin., Raspberry. Young plantations, especially near Boldrewood, 339.

RUBUS FRUTICOSUS, Aut., Common Bramble, 340.

RUBUS SUBERECTUS, And., Red-fruited Bramble. Wootton Enclosures, where it was first pointed out to me in 1859 by Mr. Lees, 340 (3).

ROSA SPINOSISSIMA, Lin., Burnet-leaved Rose. Not uncommon round Ashley and Wootton, 341.

CRATÆGUS OXYACANTHA, Lin., Common Hawthorn, 360.

PYRUS MALUS, Lin., Wild Crab, 363.

PYRUS TORMINALIS, Sm., Wild Service Tree, 364.

PYRUS ARIA, Sm., White Beam, 365.

PYRUS AUCUPARIA, Gaert., Mountain Ash. Probably naturalized, 366.

EPILOBIUM MONTANUM, Lin., Mountain Willow Herb, 370.

Isnardia palustris, Lin., Marsh Isnardia. Found at Brockenhurst by Mr. Borrer; *Phytologist*, vol. iii. p. 368. *See* also iv. p. 754; 376.

CIRCÆA LUTETIANIA, Lin., Enchanter's Nightshade. In most of the old woods, 377.

LYTHRUM SALICARIA, Lin., Purple Willow Herb. The Forest pools, 390.

TILLÆA MUSCOSA, Lin., Moss-like Tillæa, Everywhere in the Forest, 407.

HEDERA HELIX, Lin., Common Ivy, 438.

CORNUS SANGUINEA, Lin., Cornel-tree, 439.

HYDROCOTYLE VULGARIS, Lin., Marsh Pennywort. Throughout the Forest, 441.

SANICULA EUROPÆA, Lin., Wood Sanicle. In most of the old woods, 442.

VISCUM ALBUM, Lin., Mistletoe. Grows chiefly on the black poplar, especially near Godshill. I have never seen it on the oak. Abundance of it may be found in the apple-trees in the Forest keeper's garden at Boldrewood, 503.

SAMBUCUS NIGRA, Lin., Common Elder, 504.

SAMBUCUS EBULUS, Lin., Danewort. "Near Lyndhurst," T. B. Rake, Esq., 505.

VIBURNUM OPULUS, Lin., Guelder Rose, 506.

LONICERA PERICLYMENUM, Lin., Common Honeysuckle, 508.

GALIUM VERUM, Lin., Ladies' Bedstraw, 513.

HIERACIUM VULGATUM, Freis., Wood Hawkweed, 568 (24).

SERRATULA TINCTORIA, Lin., Sawwort. Throughout the Forest, 594,

CARDUUS MARIANUS, Lin., Blessed Thistle. Forest roadsides, 598.

CARDUUS PRATENSIS, Huds., Meadow Thistle. Abundant in the southern part of the Forest round Wootton, 604.

BIDENS CERNUA, Lin., Nodding Bur Marigold. Waste lands round and in the Forest. Has a fine effect on the landscape near Godshill; common, however, throughout the district, 617.

EUPATORIUM CANNABINUM, Lin., Hemp Agrimony. Gives a rich appearance to the Forest streams; but, of course, abundant elsewhere, 619.

FILAGO MINIMA, Fries. The Least Cudweed, 634.

SOLIDAGO VIRGAUREA, Lin., Golden Rod. Throughout the

Forest, 642.

Senecio sylvaticus, Lin., Wild Groundsel. This plant, with the common nettle, is especially remarkable in the Forest, as an indication of the former existence of habitations. It may be noticed in Sloden, Eyeworth, and Island's Thorn, near the Romano-British potteries. (*See* ch. xviii. p. 216, foot-note.)

ACHILLEA PTARMICA, Lin., Sneesewort. Throughout the Forest, 671.

CAMPANULA ROTUNDIFOLIA, Lin., Nodding-flowered Harebell, 675.

JASIONE MONTANA, Lin., Sheep's-bit Scabious, 687.

ERICA TETRALIX, Lin., Cross-leaved Heath, 690.

ERICA CINEREA, Lin., Fine-leaved Heath, 692.

CALLUNA VULGARIS, Salisb., Common Ling, 695.

VACCINIUM MYRTILLUS, Lin., the Bilberry; better known in the Forest as the "Blackheart," 703.

ILEX AQUIFOLIUM, Lin., Common Holly. Most abundant, 713.

FRAXINUS EXCELSIOR, Lin., Common Ash. Scarce, 715.

VINCA MINOR, Lin., Lesser Periwinkle. Hedges round and in the Forest, as at Sway, Ashley, Canterton, 716.

GENTIANA PNEUMONANTHE, Lin., Calathian Violet. Very plentiful some years at Wootton, 719.

CICENDIA FILIFORMIS, Reich., Least Gentianella. Damp places in the Forest. Rev. H. M. Wilkinson gives especially the neighbourhood of Burley, 723.

MENYANTHES TRIFOLIATA, Lin., Common in most of the Forest pools on the South, 727.

CUSCUTA EPITHYMUM, Sm., Lesser Dodder. Distributed through the Forest, on the heath and furze, 734.

VERBASCUM BLATTARIA, Lin., Moth Mullein. Not common in the Forest. I have seen a few specimens on Ashley Common; but, in 1861, a field near the new parsonage was covered with it and the viper's bugloss. Mr. Rake has found it growing at Gorely, on the north-west side of the Forest, 744.

VERONICA SCUTELLATA, Lin., Narrow-leaved Marsh Speedwell. Not common. Mr. Wilkinson gives marshy spots near Sandford and Crow, on the borders of the Forest, 758.

EUPHRASIA OFFICINALIS, Lin., Common Eyebright, 766.

MELAMPYRUM PRATENSE, Lin., Meadow Cow Wheat, 770.

PEDICULARIS PALUSTRIS, Lin., Marsh Lousewort, 772.

PEDICULARIS SYLVATICA, Lin., Common Lousewort, 778.

DIGITALIS PURPUREA, Lin., Purple Foxglove, 778.

LIMOSELLA AQUATICA, Lin., Common Mudwort. Found by Mr. H. C. Watson on the road from Brockenhurst to Lyndhurst, after you pass the bridge from the former place, 788.

OROBANCHE MAJOR, Angl., Great Broom-rape. On the furze, especially in the northern parts of the Forest, 790.

MENTHA AQUATICA, Lin., Water Mint, but of course throughout the district, 806.

MENTHA PRATENSIS, Sole, Meadow Mint. I give this on the authority of Sole, quoted by Dawson Turner, as found in the Forest, 807 e.

MENTHA PULEGIUM, Lin., Pennyroyal. Not uncommon, especially in wet places on the southern parts of the Forest, round Wilverley and Holmsley, 809.

THYMUS SERPYLLUM, Lin., Wild Thyme, 810.

CALAMINTHA CLINOPODIUM, Spen., Wild Basil, 815.

MELITTIS MELISSOPHYLLUM, Lin., Bastard Balm. Very plentiful on the outer bank of Wootton Enclosure, looking westward, 817.

TEUCRIUM SCORODONIA, Lin., Wood Sage, 818.

STACHYS BETONICA, Benth., Wood Betony, 836.

SCUTELLARIA MINOR, Lin., Lesser Skull-cap. Damp places in the Forest, especially round Wootton, 846.

Pulmonaria angustifolia, Lin., Narrow-leaved Lungwort. Very common round Wootton, both with and without spots on the leaves. (*See* Watson's *New Botanist's Guide*, vol. ii., p. 569; and the *Cybele Britannica*, vol. iii., p. 488), 868.

Pinguicula Lusitanica, Lin., Pale Butterwort. Bogs round Wootton; Ashley Common, where the Rev. P. Somerville first pointed it out to me. Mr. Wilkinson also gives Sandford and Crow as localities, 874.

Utricularia vulgaris, Lin., Water Milfoil. Pools in the southern part of the Forest, as also on Ashley Common, 875.

Utricularia minor, Lin., Smaller Bladderwort. Hinchclsca Bog, where I found it in 1859, with Mr. Lees. The Rev. H. M. Wilkinson gives also ponds near Burley, and Mr. Somerville ponds at the Osmanby Ford stream, 877.

Primula vulgaris, Huds., Common Primrose, 878.

Lysimachia nemorum, Lin., Wood Loosetrife, 889.

Anagallis tenella, Lin., Bog Pimpernel. In all the boggy places, 891.

Centunculus minimus, Lin., Chaff-weed, 892.

Samolus Valerandi, Lin., Brook-weed. Found it, with Mr. Lees, on Ashley Common, June 14, 1859. "The Beaulieu River," Rev. H. M. Wilkinson. It shows a decided partiality for the southern part towards the sea, 893.

Littorella lacustris, Lin., Common Shore-weed, 905.

Euphorbia amygdaloides, Lin., Wood Spurge, 974.

Mercurialis perennis, Lin., Perennial Mercury, 976.

Quercus Robur, Lin., the Oak, 988.

Quercus sessiliflora, Sm., Sessile-fruited Oak. The finest in the Forest are now in the Brook Woods, 988c.

Fagus sylvatica, Lin., the Beech, 989.

Carpinus Betulus, Lin., Hornbeam. Scarce, 990.

Corylus Avellana, Lin., Hazel, 991.

Alnus glutinosa, Lin., Common Alder, 992.

Betula alba, Lin., Common Birch, 993.

Populus alba, Lin., White Poplar, 995.

Populus tremula, Lin., Aspen, 997.

Populus nigra, Lin., Black Poplar, 998.

Salix viminalis, Lin., Common Osier, 1007.

Salix repens, Lin., Creeping Willow, 1017.

Myrica Gale, Lin., Bog Myrtle. The "Gold Withy" of the Forest, 1023.

Spiranthes autumnalis, Rich., Late-flowering Lady's Tresses. Very common in the pastures near the Forest, and on the turfy spots of the Forest lanes on the southern part, 1033.

Spiranthes æstivalis, Rich., Early-flowering Lady's Tresses. Found by Bromfield and Mr. Bennett in bogs near Lyndhurst toll-gate. *Phytologist*, vol. iii. p. 909; iv. p. 754; 1034.

Epipactis latifolia, Sm. Chewton Glen and woods running into the Forest. The Rev. P. Somerville also gives Ashley Common, 1039.

Orchis latifolia, Lin., Broad-leaved Meadow Orchis. Hinchelsea Bog. Mr. Wilkinson also gives the neighbourhood of Burley, 1052.

Gymnadenia Conopsea, Br., Fragrant-scented Orchis. Very plentiful on the south side of the railway, between Burley and Batson's Clump, about a quarter of a mile above the large "Shade pond." To be found also between Bushy Bratley and Boldrewood, 1054.

Habenaria bifolia, Br. Common in most of the open parts of the Forest, 1055.

Gladiolus Illyricus, Koch. First discovered in the Forest by the Rev. W. H. Lucas. (*See Phytologist*, Sept., 1857.) Road from Boldrewood to Lyndhurst; path from Liney Hill Wood to Rhinefield; Oakley Plantation, near Boldrewood; and the neighbourhood of the Knyghtwood Oak, where Mr. Rake and myself saw it in great abundance, July 11, 1862. In all these localities it is confined to the light sand, growing especially amongst the common brake, and seldom, if ever, extends into the heather, which grows close round. On some specimens which I forwarded, Mr. Watson observes, in speaking of the distinction between *Gladiolus imbricatus* and *illyricus*:—"The New Forest plant has the obovate capsules, hardly so much keeled, however, as described by French botanists, unless the keel becomes sharper with advancing age."

NARCISSUS PSEUDO-NARCISSUS, Lin., Daffodil. South side of the Forest near Wootton, 1073.

HYACINTHUS NON-SCRIPTUS, Lin., Bluebell, 1093.

RUSCUS ACULEATUS, Lin., Butcher's Broom. The "Kneeholm" of the Forest, 1097.

HYDROCHARIS MORSUS-RANÆ, Lin., Common Frog-bit, 1107.

ALISMA RANUNCULOIDES, Lin. Ashley and Chewton Commons. Pulteney gives "Sopley, near the Avon," 1110.

ACTINOCARPUS DAMASONIUM, Br., Star-headed Water Plantain. "Barton Common," the Rev. P. Somerville, 1112.

POTAMOGETON PLANTAGINEUS, Ducroz., Plantain-leaved Pond-weed. Boggy streams, 1134.

TYPHA LATIFOLIA, Lin., Reed-mace, 1147.

TYPHA ANGUSTIFOLIA, Lin., Lesser Reed-mace. Ponds at Wootton, 1148.

JUNCUS SQUARROSUS, Lin., Moss-rush Goose-corn, 1163.

LUZULA SYLVATICA, Bich., Great Wood Rush, 1169.

LUZULA PILOSA, Willd., Broad-leaved Hairy Wood Rush, 1170.

NABTHECIUM OSSIFRAGUM, Huds., Lancashire Bog Asphodel, 1175.

SCILÆNUS NIGRICANS, Lin., Black Bog Rush. Bogs round Holmsley, 1179.

RHYNCHOSPORA ALBA, Vahl., White Beak Rush, 1180.

RHYNCOSPORA FUSCA, Sm., Brown Beak Rush. Valley of the Osmanby Ford stream, Rev. H. M. Wilkinson, 1181.

SCIRPUS SETACEUS, Lin., Bristle-stalked Club Rush, 1186.

SCIRPUS CÆSPITOSUS, Lin., Scaly-stalked Club Rush, 1196.

SCIRPUS FLUITANS, Lin., Floating Club Rush, 1198.

ERIOPHORUM ANGUSTIFOLIUM, Rh., Common Cotton Grass, 1200.

CAREX PULICARIS, Lin., Flea Carex, 1205.

CAREX STELLULATA, Good., Little Prickly Carex, 1209.

CAREX OVALIS, Good., Oval-spiked Carex, 1211.

CAREX REMOTA, Lin., Remote Carex. The Forest streams, 1214.

CAREX INTERMEDIA, Good., Soft brown Carex. Boggy places, 1217.

CAREX ARENARIA, Lin., Sea Carex. The south side of the Forest, towards the sea.

CAREX DIVULSA, Good., Grey Carex, 1221.

CAREX VULPINA, Lin., Great Compound Prickly Carex, 1222.

CAREX FLAVA, Lin., Yellow Carex, 1234.

CAREX FULVA, Good., Tawny Carex, 1249.

CAREX PANICEA Lin., Pink-leaved Carex, 1241.

CAREX SYLVATICA, Huds., Pendulous Wood-Carex, 1247.

CAREX PSEUDO-CYPERUS, Lin., Cyperus-like Carex, 1249.

CAREX GLAUCA, Scop., Glaucous Heath Carex, 1250.

CAREX HIRTA, Lin., Hairy Carex,1257.

CAREX PALUDOSA, Good., Lesser Common Carex, 1260.

CAREX RIPARIA, Curtis., Great Common Carex, 1261.

PHALARIS ARUNDINACEA, Lin., Reed Canary Grass, 1269.

AGROSTIS SETACEA, Curtis., Bristleleaved Bent Grass. Broomy and Bratley. "Near Lymington," Turner, 1289.

AGROSTIS CANINA, Lin., Brown Bent Grass, 1290.

AGROSTIS VULGARIS, With., Common Bent Grass, 1291.

AGROSTIS ALBA, Lin., Marsh Bent Grass, 1292.

ARUNDO CALAMAGROSTIS. Lin., Purple-flowered Small Reed. "Near Marchwood," W. A. Bromfield, 1295.

AIRA CÆSPITOSA, Lin., Turfy Hair Grass, 1300.

AIRA FLEXUOSA, Lin., Wavy Hair Grass, 1302.

Aira caryophyllea, Lin., Silver Grass, 1303.

Aira præcox, Lin., Early Hair Grass, 1304.

Triodia decumbens, Beauv., Decumbent Heath Grass, 1315.

Molinia cœrulea, Mcench., Heath Purple Melic Grass, 1319.

Festuca bromoides, Lin., Barren Fescue Grass, 1341.

Festuca ovina, Lin., Sheep's Fescue Grass, 1342.

Festuca rubra, Lin., Creeping Fescue Grass, 1344.

Nardus stricta, Lin., Common Mat Grass, 1370.

Pilulifera globulifera, Lin., Pilwort or Peppergrass. Bogs round Holmsley. 1419.

Equisetum limosum, Lin., Smooth Naked Horsetail, 1425.

Proceeding now to the plants of the Valley of the Avon, and the cultivated districts round Christchurch, and Lymington, and Beaulieu, we shall be able to see those colonists which follow the footsteps of man, the pascual flowers of the meadows, and the Flora of the Avon. Where not particularly named, the plants are in many cases found also distributed in the Forest; but, being on the whole more characteristic of the Valley, are therefore inserted in this list.

Thalictrum flavum, Lin., Common Meadow Rue, 4.

Adonis autumnalis, Lin., Common Pheasant Eye. Mudeford, 9.

Myosurus minimus, Lin, Least Mouse-tail. Cornfields round Milton, 10.

Ranunculus hederaceus, Lin., Ivy-leaved Crowfoot, 13.

Ranunculus Ficaria, Lin., Common Pilewort, 14.

Ranunculus Flammula, Lin., Lesser Spearwort, 15.

Ranunculus Lingua, Lin., Greater Spearwort. Used to be very common on Ashley Common (now enclosed), growing in the pools with *Osmunda regalis*, 16.

Ranunculus acris, Lin., Upright Meadow Crowfoot, 19.

Ranunculus repens, Lin., Creeping Crowfoot, 20.

RANUNCULUS BULBOSUS, Lin., Bulbous Crowfoot, 21.

RANUNCULUS SCELERATUS, Lin., Celery-leaved Crowfoot, 23.

RANUNCULUS PARVIFLORUS, Lin., Small-flowered Crowfoot. "Hedgebanks between Bisterne and Ringwood," Rev. H. M. Wilkinson. Ray and Bromfield give Lymington, 24.

PAPAVER ARGEMONE, Lin., Longheaded Rough Poppy, 40.

PAPAVER DUNIUM, Lin., Long Smooth-headed Poppy, 41.

PAPAVER RHÆAS, Lin., Field Poppy, 42.

CHELIDONIUM MAJUS, Lin., Common Celandine, 45.

CORYDALIS CLAVICULATA, D. C, White-flowered Fumitory, 48.

FUMABIA CAPREOLATA, Lin., Ramping Fumitory, 50.

FUMARIA OFFICINALIS, Lin., Common Fumitory, 51.

CORONOPUS RUELLII. Lin., Common Wart-Cress, 58.

THLASPI ARVENSE, Lin., Penny-Cress, 60.

CAPSELLA BURSA PASTORIS, D. C. Shepherd's Purse, 63.

LEPIDIUM SMITHII, Hook., Smith's Peppermint, 69.

LEPIDIUM CAMPESTRE, Br., Field Mustard, 70.

DRABA VERNA, Lin., Common Whitlow Grass, 79.

CARDAMINE PRATENSIS, Lin., Lady's Smock, 85.

CARDAMINE HIRSUTA, Lin., Hairy Marsh Butter-Cress, 86.

ARABIS THALIANA, Lin., Common Thale-Cress, 88.

BARRAREA VULGARIS, Br., Common Winter Cress, 95.

BARRAREA PRÆCOX, Br., Early Winter or American Cress. "Grows on the bridge at Christchurch," Rev. H. M. Wilkinson, 97.

NASTURTIUM OFFICINALE, Br., Water Cress, 98.

NASTURTIUM TERRESTRE, Br., Land Cress, 99.

SISYMRRIUM OFFICINALE, Scop., Common Hedge Mustard, 102.

ERYSIMUM ALLIARIA, Lin., Hedge Garlic, 107.

CHEIRANTHUS CHEIRI, Lin., Wallflower. Walls of the Priory Church, Christchurch, 109.

SINAPIS ARVENSIS, Lin., Wild Mustard, 116.

RAPHANUS RHAPHANISTRUM, Lin., Wild Radish, 123.

HELIANTHEMUM VULGARE, Gasrt., Common Rock Rose, 128.

VIOLA TRICOLOR, Lin., Heartsease, 136.

Viola hirta, Lin., Hairy Violet. "Grows at Bisterne," Rev. H. M. Wilkinson. On the specific distinctions between this and the next, see what my friend the late Mr. Cheshire said in the *Phytologist.*

VIOLA ODORATA, Lin., Sweet Violet, 135.

DIANTHUS PLUMARIUS, Lin., Feathered Pink. Cloister walls, Beaulieu, 147.

DIANTHUS ARMERIA, Lin., Debtford Pink. First discovered by T. B. Rake, Esq., on a bank in a lane near the Hucklebrook, Fordingbridge, where I saw it, with him, growing, June, 1862.

SAPONARIA OFFICINALIS, Lin., Common Soapwort. The Christchurch and Ringwood Road, near the latter place; Bashley, 151.

SILENE INFLATA.Sm., Bladder Catchfly, 152.

SILENE ANGLICA, Lin., English Catchfly, 155.

LYCHNIS FLOS-CUCULI, Lin., Ragged Robin, 162.

LYCHNIS DIURNA, Sibth., Red Campion, 163.

LYCHNIS VESPERTINA, Sibth., White Campion, 164.

LYCHNIS GITHAGO, Lam., Corn Cockle, 165.

SAGINA PROCUMBENS, Lin., Procumbent Pearlwort. 167.

SAGINA APETALA, Lin., Erect Pearlwort, 169.

SAGINA NODOSA, Lin., Knotted Spurrey, 171.

SPERGULA ARVENSIS, Lin., Common Spurrey, 172.

ARENARIA SERPYLLIFOLIA, Lin., Thyme-leaved Sandwort, 178.

ARENARIA TRINERVIS, Lin., Plantain-leaved Sandwort, 182.

STELLARIA MEDIA, With., Common Chickweed, 185.

STELLARIA HOLOSTEA, Lin., Greater Stitchwort, 186.

STELLARIA GLAUCA, With., Glaucous Stitchwort, 187.

STELLARIA GRAMINEA, Lin., Grassyleaved Stitchwort, 188.

STELLARIA ULIGINOSA, Murr., Bog Stitchwort, 189.

CERASTIUM AQUATICUM, Lin., Water Chickweed, 191.

CERASTIUM GLOMERATUM, Thail., Broad-leaved Mouse-ear Chickweed, 192.

CERASTIUM TRIVIALE, Link., Narrow-leaved Mouse-ear Chickweed, 193.

LINUM CATHARTICUM, Lin., Purging Flax, 202.

MALVA MOSCHATA, Lin., Musk Mallow. Lanes near the Forest, 204.

MALVA SYLVESTRIS, Lin., Common Mallow, 205.

MALVA ROTUNDIFOLIA, Lin., Round-leafed Dwarf Mallow, 206.

ERODIUM CICUTARIUM, Sin., Hemlock Stork's-bill, 228.

GERANIUM PUSILLUM, Lin., Small-flowered Crane's-bill, 234.

GERANIUM MOLLE, Lin., Common Dove's-foot Crane's-bill, 235.

GERANIUM DISSECTUM, Lin., Jagged-leaved Crane's-bill, 236.

GERANIUM COLUMBINUM, Lin., Long-stalked Crane's-bill, 237.

GERANIUM LUCIDUM, Lin., Shining-leaved Common Crane's-bill. In the neighbourhood of Fordingbridge, Mr. Rake found it growing abundantly, June 17, 1862, the only station of which I am aware.

GERANIUM ROBERTIANUM, Lin., Herb Robert, 239.

ONONIS ARVENSIS, Lin., Rest Harrow, 254.

MEDICAGO SATIVA, Lin., Lucern, 258.

MEDICAGO LUPULINA, Lin., Black Medick, 260.

MELILOTUS OFFICINALIS, Willd., Common Melilot, 264.

TRIFOLIUM REPENS, Lin., White Trefoil, 267.

TRIFOLIUM SUBTERRANEUM, Lin. "Gravelly pastures at Bisterne." Rev. H. M. Wilkinson. 268.

TRIFOLIUM PRATENSE, Lin., Purple Clover, 271.

TRIFOLIUM ARVENSE, Lin., Hare's-foot Trefoil, 275.
TRIFOLIUM PROCUMBENS, Lin., Hop Trefoil, 281.
TRIFOLIUM MINUS, Relh., Lesser Yellow Trefoil, 282.
LOTUS MAJOR, Scop., Large Bird's-foot Trefoil, 284.
LOTUS CORNICULATUS, Lin., Common Bird's-foot Trefoil, 283.
ORNITHOPUS PERPUSILLUS, Lin., Bird's-foot, 291.
VICIA CRACCA, Lin., Tufted Vetch, 297.
VICIA SATIVA, Lin., Common Vetch, 298.
VICIA SEPIUM, Lin., Bush Vetch, 301.
VICIA HIRSUTA, Koch., Hairy-podded Tare, 303.
VICIA TETRASPERMA, Koch., Smooth-podded Tare, 304.
LATHYRUS PRATENSIS, Lin., Meadow Vetchling, 308.
SPIRÆA ULMARIA, Lin., Meadowsweet, 317.
SPIRÆA SALICIFOLIA, Ian., Willow-leaved Spiriea. "Grows near Bisterne, hut perhaps not truly wild," Rev. H. M. Wilkinson, 319.
GEUM URBANUM, Lin., Herb Bennet, 321.
GEUM RIVALE, Lin., Water Avens, 322.
AGRIMONIA EUPATORIA, Lin., Common Agrimony, 323.
POTENTILLA ANSERINA, Lin., Silverweed, 327.
POTENTILLA ARGENTEA, Lin., Horny Cinquefoil. Sandy fields in the neighbourhood of the Forest, 328.
POTENTILLA REPTANS, Lin., Creeping Cinquefoil, 331.
POTENTILLA FRAGARIASTRUM, Eh., Barren Strawberry, 333.
RUBUS CORYLIFOLIUS, Sm., Hazel-leaved Bramble, 340 (36).
RUBUS CÆSIUS, Lin., Dewberry, 340 (38).
ROSA CARINA, Lin., Dog-rose, 351.
ROSA ARVENSIS, Lin., Trailing Dogrose, 353.
POTERIUM SANGUISORBA, Lin., Common Salad Burnet, 355.
ALCHEMILLA ARVENSIS, Lam., Parsley Piert, 358.
EPILOBIUM ANGUSTIFOLIUM, Lin., French Willow-Herb, 367.
EPILOBIUM HIRSUTUM, Lin., Great Hairy Willow-Herb, 368.
EPILOBIUM PARVIFLORUM, Schreb., Small-flowered Willow-

Herb, 369.

EPILOBIUM PALUSTRE, Lin, Marsh Willow-Herb, 372.

EPILOBIUM TETRAGONUM, Lin., Square-stalked Willow-Herb (?), 373.

MYRIOPHYLLUM VERTICILLATUM, Lin., Whorl-flowered Water Milfoil, "Sopley," Garnier and Poulter, Rev. H. M. Wilkinson, 380.

MYRIOPHYLLUM SPICATUM, Lin., Spiked Water Milfoil, 381.

CALLITRICHE VERNA, Lin., Vernal Water Starwort, 383.

PEPLIS PORTULA, Liu., Water Purslane, 391.

BRYONIA DIOICA, Lin., White Briony, 393.

MONTIA FONTANA, Lin., Water Blinks, 394.

Claytonia perfoliata, Don., American Salad. First discovered by Mr. Hussey near Mudeford (*see Phytologist*, N. S. vol. i. p. 389). I received specimens from Dr. Stevens from the same place, May 11th, 1862.

SCLERANTHUS ANNUUS, Lin., Annual Knawel, 399.

RIBES RUBRUM, Lin., Red Currant. With the next "at Bisterne, apparently wild," Rev. H. M. Wilkinson, 404.

RIBES GROSSULARIA, Lin., Gooseberry, 406.

SEDUM TELEPHIUM, Lin., Everlasting Orpine, 409.

SEDUM ANGLICUM, Huds., English Stonecrop. "Avon Tyrrell," Rev. H. M. Wilkinson, 412.

SEDUM ACRE, Lin., Biting Stonecrop, 414.

Sedum reflexum, Lin., Crooked Yellow Stonecrop. This is only a casual escape. And, perhaps, like *Sempervivum tectorum*, ought to be excluded. *See* Bromfield in *Phytologist*, vol. iii. pp. 372, 416.

Cotyledon Umbilicus, Lin., Common Navelwort. "Road from Redbridge into the New Forest," W. Pamplin, quoted in Watson's *New Botanist's Guide*; "Dragon Lane, Bisterne," Rev. H. M. Wilkinson, 418.

SAXIFRAGA TRIDACTYLITES, Lin., Rue-leaved Saxifrage, 430.

CHRYSOPLENIUM OPPOSITIFOLIUM, Lin., Opposite-leaved

Golden Saxifrage, 434.

ADOXA MOSCHATELLINA, Lin., Tuberous Moschatel, 437.

CONIUM MACULATUM, Lin., Common Hemlock, 446.

HELOSCIADIUM NODIFLOBUM, Koch., Procumbent Marshwort, 454.

HELOSCIADIUM INUNDATUM, Koch., Least Water Marshwort, 455.

ÆGOPODIUM PODAGRARIA, Lin., Common Gout-weed, 457.

BUNIUM FLEXUOSUM, With., Earth-Nut, 461.

PIMPINELLA SAXIFRAGA, Lin., Common Burnet Saxifrage, 462.

SIUM ANGUSTIFOLIUM, Lin., Narrow-leaved Water-Parsnep, 465.

ŒNANTHE FISTULOSA, Lin., Common Water-Dropwort, 470.

Œnanthe pimpinelloides, Lin., Parsley Water-Dropwort. Plentiful round Milford in 1859. *See* Bromfield in *Phytologist*, vol. iii. p. 405. 471.

ŒNANTHE CROCATA, Lin., Water Hemlock, 473.

ŒNANTHE PHELLANDRIUM, Lin., Fine-leaved Water-Dropwort, 474.

ÆTHUSA CYNAPIUM, Lin., Fools' Parsley, 475.

ANGELICA SYLVESTRIS, Lin., Wild Angelica, 482.

HERACLEUM SPHONDYLIUM, Lin., Cow Parsnep, 487.

DAUCUS CAROTA, Lin., Common Carrot, 489.

TORILIS ANTHRISCUS, Gærtn., Upright Hedge Parsley, 493.

TORILIS INFESTA, Spr., Spreading Hedge Parsley, 494.

TORILIS NODOSA, Gaertn., Knotted-Hedge Parsley, 495.

SCANDIX PECTEN, Lin., Shepherd's Needle, 496.

ANTHRISCUS VULGARIS, Pers., Common Beaked Parsley, 497.

ANTHRISCUS SYLVESTRIS, Hoff., Wild Chervil, 498.

CILEROPHYLLUM TEMULENTUM, Lin., Hare's Parsley, 500.

GALIUM PALUSTRE, Lin., Marsh Goose-Grass, 515.

GALIUM ULIGINOSUM, Lin., Rough Marsh Bed-Straw, 516.

GALIUM SAXATILE, Lin., Mountain Bed-Straw. Perhaps this ought rather to come under the head of Forest Plants, 517.

GALIUM MOLLUGO, Lin., Great Hedge Bed-Straw, 519.

GALIUM APARINE, Lin., Goose-Grass, 523.

SHERARDIA ARVENSIS, Lin., Field Madder.

ASPERULA ODORATA, Lin., Scented Woodruff, 527.

VALERIANA DIOICA, Lin., Marsh Valerian, 531.

VALERIANA OFFICINALIS, Lin., Common Valerian, 532.

FEDIA OLITORIA, Vahl., Lamb's Lettuce, 534.

FEDIA DENTATA, Bieb., Oval-fruited Corn Salad, 537.

DIPSACUS SYLVESTRIS, Lin., Teasel, 539.

DIPSACUS PILOSUS, Lin., Shepherd's Rod. "Woods near Hale," T. Beaven Rake, 540.

SCABIOSA SUCCISA, Lin., Devil's-bit Scabious, 541.

KNAUTIA ARVENSIS, Coult., Field Scabious, 543.

TRAGOPOGON PRATENSIS, Lin., Meadow Goat's Beard, 544.

HELMINTHIA ECHIOIDES, Gærtn., Echium-like Ox-tongue. Efford Mill. Pennington, 546.

THRINCIA HIRTA, Roth., Rough Thrincia, 548.

APARGIA HISPIDA, Willd., Rough Hawkbit, 549.

APARGIA AUTUMNALIS, Willd., Autumnal Hawkbit, 550.

HYPOCHÆRIS RADICATA, Lin., Long-rooted Cat's-ear, 553.

LACTUCA MURALIS, Less., Wall Lettuce. Beaulieu and Ellingham churches, 557.

SONCHUS ARYENSIS, Lin., Field Sow Thistle, 559.

SONCHUS ASPER, Hoffm., Rough Sow Thistle, 560.

SONCHUS OLERACEUS, Lin., Sow Thistle, 561.

CREPIS VIRENS, Lin., Smooth Crepis, 563.

HIERACIUM PILOSELLA, Lin., Mouse-ear Hawkweed, 568.

HIERACIUM UMBELLATUM, Lin., Narrow-leaved Hawkweed. "Bisterne," Rev. H. M. Wilkinson.

HIERACIUM BOREALE, Fries., "Bisterne," Rev. H. M. Wilkinson.

TARAXACUM OFFICINALE, Wigg., Dandelion, 588.

Lapsana communis, Lin., Nipplewort, 590.

Cichorium intybus, Lin., Common Chicory, 591.

Arctium Lappa, Lin., Burdock, 592.

Carduus nutans, Lin., Nodding Thistle. Roadsides round the Forest, 395.

Carduus lanceolatus, Lin., Spear Thistle, 599.

Carduus palustris, Lin., Marsh Thistle, 601.

Carduus arvensis, Lin., Field Thistle, 602.

Carduus acaulis, Lin., Dwarf Thistle, 606.

Onopordum acanthium, Lin., Cotton Thistle, 608.

Carlina vulgaris, Lin., Common Carline Thistle, 609.

Centaurea nigra, Lin., Black Knapweed, 611.

Centaurea Cyanus, Lin., Cornflower, 612.

Centaurea Scabiosa, Lin., Great Knapweed, 613.

Bidens tripartita, Three-lobed Bur-marigold, 618.

Tanacetum vulgare, Lin., Common Tansy, 622.

Artemisia vulgaris, Lin., Mugwort, 626.

Gnaphalium sylvaticum, Lin., Upright Cudweed, 630.

Gnaphalium uliginosum, Lin., Marsh Cudweed, 632.

Filago germanica, Lin., Erect Cudweed. 635.

Tussilago Farfara, Lin., Common Coltsfoot, 6.37.

Erigeron acris, Lin., Blue Fleabane. "Near Milton," Rev. P. Somerville, 639.

Senecio vulgaris, Lin., Groundsel, 643.

Senecio Jacobæa, Lin., Ragwort, 648.

Senecio aquaticus, Huds., Water Ragwort. "Bisterne," Rev. H. M. Wilkinson, 648*.

Inula Conyza, D. C, Ploughman's Spikenard. Sopley, Rev. H. M. Wilkinson, 656.

Pulicaria dysenterica, Gærtn., Common Fleabane, 658.

Pulicabia vulgaris, Gærtn., Small Fleabane. "Marchwood," W. A. Bromfield (*Phytologist*, vol. iii. p. 433); "Bisterne," Rev. H. M. Wilkinson, 659.

Bellis perennis, Lin, Common Daisy, 660.

Chrysanthemum segetum, Lin., Corn Marigold, 661.

Chrysanthemum Leucanthemum, Lin., Great Ox-eye Daisy, 662.

Pyrethrum Parthenium, Sm., Feverfew, 663.

Pyrethrum inodorum, Sm., Scentless Feverfew, 664.

Matricaria Chamomilla, Lin., Wild Chamomile, 665.

Anthemis nobilis, Lin., Common Chamomile, 666.

Anthemis arvensis, Lin., Corn Chamomile, 668.

Anthemis Cotula, Lin., Stinking Mayweed, 669.

Achillea Millefolium, Lin, Yarrow, 6T2.

Campanula patula, Lin., Spreading Bell-flower. Avon Tyrrell, 676.

Specularia hybrida, D. C., Venus's Looking-glass. "Corn-fields near Sandford," Rev. H. M. Wilkinson, 684.

Ligustrum vulgare, Lin., Common Privet, 714.

Erythræa Centaurium, Pers., Common Centaury, 724.

Chlora perfoliata, Lin., Perfoliate Yellow-wort, 725.

Convolvulus arvensis, Lin., Small Bindweed, 729.

Convolvulus sepium, Lin., Great Bindweed, 730.

Hyoscyamus niger, Lin., Henbane. Roadside bank near Ibbesley, 736.

Datura Stramonium, Lin., Thorn Apple. Near Ringwood, on the Christchurch road.

Solanum nigrum, Lin., Black Nightshade, 737.

Solanum Dulcamara, Lin., Woody Nightshade, 738.

Verbascum Thapsus, Lin., Taper Moth Mullein, 740.

Verbascum nigrum, Lin., Black Moth Mullein, 743.

Veronica arvensis, Lin., Wall Speedwell, 747.

Veronica serpyllifolia, Lin., Thyme-leaved Speedwell, 750.

Veronica Anagallis, Lin., Water Speedwell, 754.

Veronica Beccabunga, Lin., Brooklime, 755.

VERONICA OFFICINALIS, Lin., Common Speedwell, 756.

VERONICA CHAMÆDRYS, Lin., Germander Speedwell, 758.

VERONICA HEDERIFOLIA, Lin., Ivy-leaved Speedwell, 759.

VERONICA AGRESTIS, Lin., Procumbent Speedwell, 760.

VERONICA BUXBAUMII, Ten., Buxbaum's Speedwell. Mr. Rake found it in abundance not far from Fordingbridge, March, 1862, 762.

BARTSIA ODONTITES, Hnds., Red Rattle, 765.

RHINANTHUS CRISTA-GALLI, Lin., Meadow Rattle, 767.

SCROPHULARIA NODOSA, Lin., Knotty-rooted Figwort, 774.

SCROPHULARIA AQUATICA, Lin., Water Figwort, 775.

ANTIRRHINUM ORONTIUM, Lin., Field Snap-dragon. Milton and Somerford, 780.

LINARIA CYMBALARIA, Mill., Wall Toad-flax, 781.

LINARIA ELATIKE, Mill., Sharp-pointed Toad-flax, 783.

LINARIA REPENS, Ait, Creeping Toad-flax. "Marchwood," Borrer, 784.

LINARIA VULGARIS, Mill., Common Toad-flax, 785.

OROBANCHE MINOR, Sutt., Lesser Broom-rape, 793.

VERBENA OFFICINALIS, Lin., Common Vervain, 798.

SALVIA VERBENACA, Lin., Wild Clary. Roads near Christchurch; keep of Christchurch Castle; Beaulieu Churchyard, 799.

LYCOPUS EUROPÆUS, Lin., Gipsywort, 801.

MENTHA SATIVA, Lin., Hairy Water Mint, 807.

MENTHA ARVENSIS, Lin., Field Mint, 808.

CALAMINTHA ACINOS, Clairv., Basil Thyme. Fernhill Lane, 812.

CALAMINTHA OFFICINALIS, Angl., Officinal Calamint. Avon Tyrrel, 814.

AJUGA REPTANS, Lin., Common Bugle, 822.

BALLOTA NIGRA, Lin., Black Horehound, 825.

LAMIUM ALBUM, Lin., White Dead Nettle, 828.

LAMIUM AMPLEXICAULE, Lin., Henbit, 830.

LAMIUM PURPUREUM, Lin., Red Henbit, 831.
GALEOPSIS TETRAHIT, Lin., Common Hemp Nettle, 834.
STACHYS PALUSTRIS, Lin., Marsh Woundwort, 837.
STACHYS SYLVATICA, Lin., Hedge Woundwort, 838.
STACHYS ARVENSIS, Lin., Field Woundwort, 840.
GLECHOMA HEDERACEA, Lin., Ground Ivy, 841.
NEPETA CATARIA, Lin., Catmint. Near Bisterne, 842.
MARRUBIUM VULGARE, Lin., Common Horehound, 843.
PRUNELLA VULGARIS, Lin., All-heal, 844.
SCUTELLARIA GALERICULATA, Lin., Common Skull-cap. Chewton Glen, Beckton Bunny, 845.
MYOSOTIS PALUSTRIS, With., Forget-me-not, 847.
MYOSOTIS CÆSPITOSA, Schultz, Marsh Mouse Ear, 849.
MYOSOTIS ARVENSIS, Hoff., Field Marsh Ear, 852.
MYOSOTIS COLLINA, Hoff., Dwarf Mouse Ear, 853
MYOSOTIS VERSICOLOR, Lehm., Yellow and Blue Mouse Ear, 854.
LITHOSPERMUM ARVENSE, Lin., Field Gromwell, 856.
SYMPHYTUM OFFICINALE, Lin., Common Comfrey, 859.
BORAGO OFFICINALIS, Lin., Common Borage, 861.
LYCOPSIS ARVENSIS, Lin., Ox-tongue, 862.
CYNOGLOSSUM OFFICINALE, Lin., Common Hound's-tongue, 866.
ECHIUM VULGARE, Lin., Viper's Bugloss, 869.
PRIMULA VERIS, Lin., Cowslip, 880.
LYSIMACHIA VULGARIS, Lin., Yellow Loosestrife, 886.
LYSIMACHIA NUMMULARIA, Lin., Moneywort, 888.
ANAGALLIS ARVENSIS, Lin., Poor Man's Weather Glass, 890.
PLANTAGO MAJOR, Lin., Greater Plantain, 901.
PLANTAGO MEDIA, Lin., Hoary Plantain. "Beaulieu, on the clay," Rev. H. M. Wilkinson.
PLANTAGO LANCEOLATA, Lin., Rib Grass, 903.
PLANTAGO CORONOPUS, Lin., Buckthorn Plantain, 905.

Chenopodium urbicum, Lin., Erect Goose-foot, 910-.

Chenopodium rubrum, Lin., Red Goose-foot, 911.

Chenopodium album, Lin., White Goose-foot, 914.

Chenopodium Bonus-henricus, Lin., Good King Henry, 917.

Atriplex hastata, Lin., Narrow-leaved Orache, 922.

Atriplex patula, Lin., Spreading Orache, 923.

Polygonum amphibium, Lin., Amphibious Persicaria, 933.

Polygonum lapathifolium, Lin., Pale-flowered Persicaria, 934.

Polygonum Persicaria, Lin., Spotted Persicaria, 935.

Polygonum Hydropiper, Lin., Biting Persicaria, 937.

Polygonum aviculare, Lin., Common Knot Grass, 938.

Polygonum Convolvulus, Lin., Black Bindweed. On the difference between this and *C. dumetorum*, *see* Dr. Bromfield in the *Phytologist*, vol. iii. p. 765.

Rumex Hydrolapathum, Huds., Great Water Dock. The Avon, 943.

Rumex crispus, Lin., Curled Dock, 944.

Rumex obtusifolius, Lin., Blunt-leaved Dock, 947.

Rumex sanguineus, Lin., Blood veined Dock, 948.

Rumex conglomeratus, Mnr., Sharp-leaved Dock, 948*.

Rumex Acetosa, Lin., Common Sorrel, 951.

Rumex Acetosella, Lin., Sheep's Sorrel, 952.

Euphorbia helioscopia, Lin., Sun Spurge, 962.

Euphorbia exigua, Lin., Dwarf Spurge. Near the coast, 971.

Euphorbia Peplus, Lin., Petty Spurge, 972.

Urtica urens, Lin., Annual Stinging Nettle, 978.

Urtica dioica, Lin., Perrennial Stinging Nettle, 979.

Parietaria officinalis, Lin., Common Pellitory. Walls of Beaulieu Abbey, 982.

Humulus lupulus, Lin., Hop, 983.

Ulmus campestris, Sm., Common Elm. Rare in the Forest,

985a.

SALIX CINEREA, Lin., Grey Sallow, 1010.

LISTERA OVATA, Br., Common Tway-blade. Meadows round Christchurch, 1038.

EPIPACTIS PALUSTRIS, Sw., Marsh Helleborine. Chewton Glen. Rare. Mr. Rake, however, has found it growing abundantly in the neighbourhood of Fordingbridge, August, 1862, 1040.

ORCHIS MORIO, Lin., Green-winged Meadow Orchis, 1045.

ORCHIS MASCULA, Lin., Early Purple Orchis, 1046.

ORCHIS MACULATA, Lin., Spotted Palmate Orchis, 1053.

IRIS PSECDACORUS, Lin., Flag Water Iris, 1067.

GALANTHUS NIVALIS, Lin., Common Snowdrop. "Bisterne, apparently wild, though it has, doubtless, at some time or another, been planted," Rev. H. M. Wilkinson, 1074.

ALLIUM VINEALE, Lin., Common Garlic, 1083.

ORNITHOGALUM UMBELLATUM, Lin., Common Star of Bethlehem. "Bisterne. Not truly wild," Rev. H. M. Wilkinson, 1090.

TAMUS COMMUNIS, Lin., Black Bryony, 1104.

ANACHARIS ALSINASTRUM, Bab., Chickweed-like American Weed. R. Stevens, Esq., found this straggler, "July 23rd, 1862, at Knapp Mill, in a ditch leading out of the Avon," 1107*

ALISMA PLANTAGO, Lin., Greater Water Plantain, 1109.

SAGITTARIA SAGITTIFOLIA, Lin., Arrow Head. The Avon, 1113.

BUTOMUS UMBELLATUS, Lin., Flowering Rush. The Avon, 1114.

TRIGLOCHIN PALUSTRE, Lin., Marsh Arrow Grass. Banks of the Avon, 1116.

POTAMOGETON DENSUS, Lin., Close-leaved Pond Weed, 1118.

POTAMOGETON CRISPUS, Lin., Curled Pond-weed. 1124.

POTAMOGETON PERFOLIATUS, Lin., Perfoliate Pond Weed, 1125.

POTAMOGETON LUCENS, Lin., Shining Pond Weed, 1126.

POTAMOGETON NATANS, Lin., Broad-leaved Pond Weed, 1132.

Zannichellia palustris, Lin., Horned Pond Weed, 1136.

Lemna minor, Lin., Lesser Duckweed, 1138.

Lemna polyrhiza, Lin., Greater Duckweed, 1140.

Lemna trisulca., Lin.. Ivy-leaved Duckweed. The Avon, 1141.

Arum maculatum, Lin., Cuckoo-pint, 1142.

Sparganium simplex, Huds., Unbranched Bur-reed, 1145.

Sparganium ramosum, Huds., Branched Bur-reed. Found it, with Mr. Lees, in ponds at Wootton, 1147.

Juncus conglomeratus, Lin., Common Rush, 1151.

Juncus effusus, Lin., Soft Rush, 1151.

Juncus glaucus, Sibth., Hard Rush, 1152.

Juncus acutiflorus, Ehrh., Sharp-flowered jointed Rush, 1156.

Juncus lamprocarpus, Ehrh., Shining-fruited jointed Rush, 1157.

Juncus supinus, Mœnch., Whorl-headed Rush, 1159.

Juncus compressus, Jacq., Round-fruited Rush, 1160.

Juncus bufonius, Lin., Toad Rush, 1162.

Luzula campestris, "Br.," Field Wood Rush, 1172.

Scirpus lacustris, Lin., Bull Rush. The Avon, 1184.

Carex paniculata, Lin., Great Panicled Carex. "Chewton Glen," Rev. H. M. Wilkinson, 1224.

Carex vulgaris, Fries., Tufted Bog Carex, 1228.

Carex pallescens, Lin., Pale Carex, 1236.

Carex præcox, Jacq., Vernal Carex, 1251.

Carex pilulifera, Lin., Reed-headed Carex, 1252.

Leersia oryzoides, Sw., Leersia. "Bisterne and Sopley," Rev. H. M. Wilkinson; "Brockenhurst," *Phytologist*, vol. iv. p. 754; 1262*.

Anthoxanthum odoratum, Lin., Sweet-scented Vernal Grass, 1271.

Phleum pratense, Lin., Meadow Timothy Grass, 1273.

Alopecurus pratensis, Lin., Meadow Fox-tail Grass. Rare

in the Forest, 1278.

ALOPECURUS GENICULATUS, Lin, Floating Fox-tail Grass, 1279.

ALOPECURUS AGRESTIS, Lin., Slender Fox-tail Grass, 1282.

ARUNDO PHRAGMITES, Lin., Common Reed, 1294.

ARUNDO EPIGEJOS, Lin., Wood Reed, 1296.

AVENA FLAVESCENS, Lin., Yellow Oat Grass, 1311.

ARRHENATHERUM AVENACEUM, Beauvois, Oat-like Grass, 1312.

HOLCUS LANATUS, Lin., Meadow Soft Grass, 1313.

HOLCUS MOLLIS, Lin., Creeping Soft Grass, 1314.

CATABROSA AQUATICA, Presl., Water Whorl Grass, 1320.

GLYCERIA AQUATICA, Sm., Reed Meadow Grass, 1321.

GLYCERIA FLUITANS, Br., Floating Sweet Grass, 1322.

POA ANNUA, Lin., Annual Meadow Grass, 1328.

POA PRATENSIS, Lin., Smooth-stalked Meadow Grass, 1331.

POA TRIVIALIS, Lin., Roughish Meadow Grass, 1332.

BRIZA MEDIA, Lin., Common Quaking Grass, 1335.

BRIZA MINOR, Lin., Small Quaking Grass. "Corn-fields round Marchwood, perhaps introduced with the grain," W. A. Bromfield, 1336.

CYNOSURUS CRISTATUS, Lin., Crested Dog's Tail Grass, 1337.

DACTYLIS GLOMERATA, Lin., Rough Cock's-foot Grass, 1339.

FESTUCA PRATENSIS, Huds., Meadow Fescue Grass, 1347.

FESTUCA LOLIACEA, Huds. "Common at Bisterne," Rev. H. M. Wilkinson, 1347 b.

BROMUS GIGANTEUS, Lin., Tall Fescue Grass, 1348.

BROMUS STERILIS, Lin., Barren Brome Grass, 1350.

BROMUS SECALINUS, Lin., Smooth Rye Brome Grass, 1354.

BROMUS MOLLIS, Lin., Soft Brome Grass, 1356.

BROMUS RACEMOSUS, Lin.(?) "Common at Bisterne," Rev. H. M. Wilkinson, 1356 b.

BRACHYPODIUM SYLVATICUM, Beauv., Slender False Brome

Grass, 1357.

TRITICUM CANINUM, Huds., Fibrous-rooted Wheat Grass, 1359.

TRITICUM REPENS, Lin., Creeping Wheat Grass, 1360.

LOLIUM PERENNE, Lin., Common Rye Grass, 1363.

HORDEUM PRATENSE, Huds., Meadow Barley, 1367.

HORDEUM MURINUM, Lin., Wall Barley, 1368.

EQUISETUM TELMATEIA, Ehrh., Great Horsetail, 1420.

EQUISETUM ARVENSE, Lin., Field Horsetail, 1422.

EQUISETUM PALUSTRE, Lin., Marsh Horsetail, 1424.

Whilst these lists were passing through the press, H. C. Watson, Esq., sent me the following additions, all noticed by himself in August, 1861, within three or four miles of Brockenhurst:—

NASTURTIUM AMPHIBIUM, Br., Great Yellow Cress, 101.

VIOLA FLAVICORNIS, Sm., Dwarf Yellow-spurred Violet, 189 h.

EPILOBIUM ROSEUM, Schreb., Pale Smooth-leaved Willow Herb, 371.

Epilobium obscurum, Schreb. (For a description of this plant, see *Phylologist*, new series, vol. ii. p. 19.) 373 b.

EUPHRASIA GRACILIS, Fr., 766 b.

Polygonum minus, Huds., Small Creeping Persicaria. (Brorafield in the *Flora Vectensis*, p. 433, mentions it as growing in the Island.) 938.

CAREX BINERVIS, Sm., Green-ribbed Carex, 1239.

BROMUS ASPER, Lin., Hairy Wood Brome Grass, 1349.

To these also may be added *Coronopus didyma*, mentioned by Bromfield (*Phytologist*, vol. iii. p. 210) as found along the coast, but which will, perhaps, be met inland.

Gladiolus Illyricus.

Footnotes

1. The numbers after a plant refer to its numerical place in the *London Catalogue*, whose nomenclature and arrangement have been followed. The English synonyms have been chiefly taken from Smith.
2. *Scirpus parvulus* (R. and S.), mentioned by Rev. G. E. Smith as growing "on a mud-flat near Lymington," is now extinct. See Watson's *Cybele Britannica*, vol. iii. p. 78; and Bromfield, in the *Phytologist*, vol. iii., 1028.

APPENDIX III.

LIST OF THE BIRDS OF THE NEW FOREST DISTRICT.

THE best plan is, perhaps, to arrange the birds in groups, and to give a short analysis of each section, so that the reader may be able to see at a glance the more characteristic as well as rarer species. We will first of all take the Residents. In making out this list I have been principally guided—with of course certain exceptions—by the rule of admitting every bird whose nest has been found upon reliable evidence, as we may be sure that for one nest which is discovered a dozen or more remain undetected.

Peregrine Falcon. (*Falco peregrinus*, Gmel.) As this bird breeds so near, both in the Isle of Wight and along the Dorsetshire coast, it may be considered as a resident. From different lists before me, ranging over several years, it appears to have been shot and trapped in the Forest at all seasons.

Merlin. (*Falco æsalon*, Gmel.) *See* Chapter XXII.

Kestrel. (*Falco tinnunculus*, Lin.) Numerous.

Sparrow Hawk. (*Falco nisus*, Lin.) More abundant than even the kestrel, especially in the southern part of the Forest.

Common Buzzard. (*Falco buteo*, Lin.) Breeds in nearly all the old woods, but is becoming scarce. *See* Chapter XXII.

Marsh Harrier. (*Circus aruginosus*, Lin.) Rare.

Hen Harrier. (*Circuscyaneus*, Lin.) *See* Chapter XXII.

This bird has become much more numerous of late. No less than six or seven pairs were, I am sorry to say, trapped last year.

Long-Eared Owl. (*Strix otus*, Lin.) Not unfrequent. I have found it nesting round Mark Ash and Boldrewood. Mr. Rake tells me that Amberwood is also a favourite breeding station.

Barn Owl. (*Strix flummea*, Lin.) Not so common as might be expected.

Tawny Owl. (*Strix aluco*, Lin.) The most common of the three. Very often this bird may be seen during the day in the Forest mobbed by thrushes and blackbirds, and taking refuge in some of the large ivy-bushes.

Missel Thrush. (*Turdus viscivorus*, Lin.) Known throughout the Forest as the "Bull thrush."

Song Thrush. (*Turdus musicus*, Lin.)

Blackbird. (*Turdus morula*, Lin.)

Robin Redbreast. (*Sylvia rubecula*, Lath.)

Stonechat. (*Sylvia rubicula*, Lath.) Mr. Rake tells me that it breeds rather plentifully round Ogdens and Frogham, about two miles from Fordingbridge. I have also had the eggs brought me from Wootton.

Dartford Warbler. (*Sylvia provincialis*, Ks. and Bl.) Is sometimes very common in the Forest, and is generally to be seen in company with the whinchat. In some years, as in 1861, it is scarce. I have its nest, with two eggs, in my collection, taken by Mr. Farren, on Lyndhurst Heath, April 29th, 1862; but it is always difficult to find, as the bird frequents, in the breeding season, the thickest part of the high furze.

Goldencrested Regulus. (*Regulus cristatus*, Koch.) Not uncommon. Known throughout the Forest as "The thumb bird."

Great Titmouse. (*Parus major*, Lin.)

Blue Titmouse. (*Parus cæruleus*, Lin.)

Cole Titmouse. (*Parus ater*, Lin) Far more common than the next.

Marsh Titmouse. (*Parus palustris*, Lin.)

Long-tailed Titmouse. (*Parus caudatus*, Lin.) Known

throughout the Forest as the "Long-tailed caffin," or "cavin."

Pied Wagtail. (*Motacilla Yarrellii*, Gould.) Partially migratory.

Grey Wagtail. (*Motacilla boarula*, Lin.) After some hesitation, I have decided to put this bird among the residents. Yarrell (vol. i., 434) mentions it breeding near Fordingbridge, close to the upper boundary of the Forest.

Meadow Pipit. (*Anthus pratensis*, Bechst.) The "Butty lark," that is, companion bird, of the New Forest; so called because it is often seen pursuing the cuckoo, which the peasant takes to be a sign of attachment instead of anger.

Rock Pipit. (*Anthus obscurus*. Keys and Bl.) Inhabits the muddy shores of the south-eastern district.

Sky Lark. (*Alauda arvensis*, Lin.)

Wood Lark. (*Alauda arborea*, Lin.) Mr. Rake found its nest on Goreley racecourse, near Fordingbridge, on the 2nd of April, 1861, with three eggs.

Common Bunting. (*Emberiza miliaria*, Lin.)

Blackheaded Bunting. (*Emberiza schœniclus*, Lin.)

Yellow Hammer. (*Emberiza citrinella*, Lin.)

Cirl Bunting. (*Emberiza cirlus*, Lin.) I have had its eggs brought to me from the neighbourhood of Wootton; and Mr. Farren found a nest with three eggs in 1861, close to the village of Brockenhurst.

Chaffinch. (*Fringilla cœlebs*, Lin.) The "Chink" of the New Forest.

House Sparrow. (*Fringilla domestica*, Lin.)

Greenfinch. (*Fringilla chloris*, Lin.)

Hawfinch. (*Fringilla coccothraustes*, Lin.) A few pair now and then certainly remain in the Forest to breed, though I have never been fortunate enough to obtain their eggs. Great quantities were killed at Burley in the spring of 1858.

Goldfinch. (*Fringilla carduelis*, Lin.)

Bullfinch. (*Loxia pyrrhula*, Lin.) Always to be seen very busy in November amongst the young buds just formed, in the cottage gardens near the Forest.

Starling. (*Sturnus vulgaris*, Lin.)

Raven. (*Corvus corax*, Lin.) Becoming very scarce. *See* Chapter XXII.,

Crow. (*Corvus corone*, Lin.)

Rook. (*Corvus frugilegus*, Lin.)

Jackdaw. (*Corvus monedula*, Lin.)

Jay. (*Corvus glandaritis*, Lin.)

Green Woodpecker. (*Picus viridis*, Lin.) "The yaffingale" and "woodnacker" of the Forest.

Spotted Woodpecker. (*Picus major*, Lin.) Both this and the next are known throughout the Forest as the "wood-pie."

Lesser-Spotted Woodpecker. (*Picus minor*, Lin.)

Creeper. (*Certhia familiaris*, Lin.) Builds in the holes of the old ash and thorn trees. *See*, however, Chapter XXII.

Wren. (*Troglodytes Europæus*, Cuv.)

Nuthatch. (*Sitta Europæa*, Lin.)

Kingfisher. (*Alcedo ispida*, Lin.) Not very common, yet it may now and then be seen at Darrat's stream, near Lyndhurst, the brook in the Queen's Bower Wood, and the Osmanby Ford river, near Wootton.

Ringdove, (*Columba palumbus*, Lin.)

Stockdove. (*Columba œnas*, Lin.) Numerous, building in the holes of the old beech-trees.

Pheasant. (*Phasianus Colchicus*, Lin.)

Black Grouse. (*Tetrao tetrix*, Lin.) Feeds on the young shoots of heather and larch, seeds of grass, blackberries and acorns, and I have seen it repeatedly perching in the hawthorns for the sake of the berries. The "heath poult" of the Forest.

Partridge. (*Perdix cinerea*, Lath.)

Lapwing. (*Vanellus cristatus*, Meyer.)

Heron. (*Ardea cinerea*, Lath.) *See* Chapter XXII.

I have known a pair lay, in one instance, at Boldrewood, as late as June 23rd.

Common Redshank. (*Totanus calidris*, Lin.) This bird is certainly a resident throughout the year. I have repeatedly put it up during the autumn in some of the swamps near Stoney Cross, more especially in the evening, when it will hover round and round, just keeping overhead, not unlike a pewit. Several nests are yearly taken. Last year Mr. Farren found one near Burley, April 4th, with a single egg, and another, May 3rd, containing four, at Bishopsditch.

Woodcock. (*Scolopax rusticola*, Lin.) Breeds in great numbers in some seasons.

Common Snipe. (*Scolopax gallinago*, Lin.) The greatest numbers occur in December, though many remain to breed not only in the "bottoms" of the Forest, but the meadows of the Avon. Mr. Rake informs me that a Sabine's snipe (*Scolopax Sabini*, Vigors), which is now generally regarded as only a melanism of this species, was shot at Picket Post, Jan., 1859. Another was shot not far from the borders of the Forest, at Heron Court, 1836.

Water Rail. (*Rallus aquaticus*, Lin.) Most common in the winter. Some few, however, breed in the valley of the Osmanby Ford stream, where I have seen a pair or two in the summer time.

Coot. (*Fulica atra*, Lin.) A straggler generally every year remains to breed on the Avon.

Mute Swan. (*Cygnus olor*, Boie.) Large numbers belonging to Lord Normanton's swannery may be always seen on the Avon, near Fordingbridge and Ibbesley.

Wild Duck. (*Anas boschas*, Lin.) Breeds, like the teal, in most of the bottoms throughout the Forest, as also in the Avon. The fowlers round Exbury say that the wigeon, too, stays to nest; but I do not know of any authenticated case. Mr. Rake has also oberved the tufted duke as late in the year as May.

Teal. (*Anas crecca*, Lin.)

Little Grebe. (*Podiceps minor*, Lath.) Known in the Forest as the di-dapper. A few breed in the Boldre Water, and, perhaps, even in the Osmanby Ford stream. Mr. Rake tells me that it breeds plentifully in the Avon, between Fordingbridge and Downton.

Guillemot. (*Uria troile*, Lath.) Locally known as the "spratter."

Razorbill. (*Alca tarda*, Lin.)

Cormorant. (*Carbo cormoranus*, Meyer.) Locally known as the "Isle of Wight parson."

Shag. (*Carbo cristatus*, Tem.)

Herring Gull. (*Larus argentatus*, Brun.) It is to be seen at all seasons with the four birds above mentioned, breeding like them in the Freshwater Cliffs of the Isle of Wight. The shag and the cormorant were the commonest birds along the south-east coast of the Forest in Gilpin's time (vol. II. pp. 172, 302, third edition), but are now becoming rare; and Mr. More, in his excellent account of the birds of the Isle of Wight, doubts whether more than one or two pairs now annually breed in the Island.

Thus the Forest possesses in all seventy-two residents. The common buzzard, the merlin, the henharrier, the three owls, and as many woodpeckers, with the nuthatch and the stockdove, well indicate its woody and heathy character. Upon comparing this with Mr. More's list of the residents of the Isle of Wight, we find that the Forest possesses fourteen more than that Island. The principal additions consist, as might be expected, of the common buzzard, black-grouse, green and great and lesser spotted woodpeckers, common snipe, and woodcock, although by the way the last, to my knowledge, breeds in the Island, as also probably the little grebe.

The summer visitors are arranged by the date of the arrival of the main body, drawn partly from Mr. Rake's and my own observations. In a few cases, as a further criterion, I have given the dates of their nesting spread over the last four years.

Chiffchaff. (*Sylvia rufa*. Lath.) Arrives about the middle and end of March. Common.

Wheatear. (*Sylvia œnanthe.* Lath.) Follows very close after the chiffchaff; but the bird is scarce.

Sand-martin. (*Hirundo riparia*, Lath.) In 1862, Mr. Rake saw some specimens near Fordingbridge on March 15th, about a week earlier than usual.

Martin. (*Hirundo urbica*, Lin.) Arrives with the sandmartin about the end of March, though sometimes both are seen a little earlier.

Swallow. (*Hirundo rustica*, Lin.)

Wryneck. (*Yunx torquilla*, Lin.) Generally to be heard about the end of March and beginning of April. Known in the Forest as the "Little Eten bird;" and from its cry the "Weet bird." Mr. Rake both heard and saw one as late as Dec. 5, 1861.

Redstart. (*Sylvia phœnicurus*, Lath.) Beginning of April.

Thicknee. (*Œdienæmus crepitans*, Tem.) It is possible that some may remain to breed.

Nightingale. (*Sylvia luscinia*, Lath.) About the middle of May their nests are mostly found in the Forest.

Cuckoo. (*Cuculus canorus*, Lin.) May 26 and June 1 are the dates when I have found its eggs placed, in one case, at Baishley, in a hedge sparrow's, and in the other, on Beaulieu Common, in a titlark's nest.

Blackcap. (*Sylvia atricapilla*, Lath.) Arrives about the beginning and middle of April.

Rat's Wagtail. (*Motacilla campestris*, Pall.) Known in the New Forest as the "Barley bird," as it appears about the time barley is sown. Probably does not breed.

Grasshopper Warbler. (*Sylvia locustella*, Lath.) Breeds in the young plantations, but is by no means common.

Sedge Warbler. (*Sylvia Phragmitis*, Bechst.) Very scarce.

Willow Wren. (*Sylvia trochilus*, Lath.) Many are to be seen about the middle and end of April in the young enclosures, where I have frequently caught the bird on its nest.

Wood Wren. (*Sylvia sibilatrix*, Bechst.) Its nests and eggs are generally found about the same time as the willow wren's.

Whitethroat. (*Sylvia cinerea*, Lath.) Common.

Lesser Whitethroat. (*Sylvia curruca*, Lath.) Not abundant.

Whinchat. (*Sylvia rubetra*, Lath.) Known throughout the Forest as the "Furze Hacker."

Tree Pipit. (*Anthus arboreus*, Bechst.) Common.

Reed Wren. (*Sylvia arundinacea*, Lath.) The five foregoing species come much about the same time, namely, the end of April, but the reed wren is excessively scarce in the Forest, and I have only once or twice heard its note in the Beaulieu river. Mr. Hart assures me that it builds on the banks of the Avon, but its nest has yet to be found.

Landrail. (*Gallinula crex*, Lath.) About the end of April or beginning of May. A good many yearly build round Milton, and the south parts of the Forest, and even in the interior, as at Fritham and Alum Green.

Common Sandpiper. (*Tetanus hypoleucos*. Tem.) A pair now and then remain to breed at Whitten pond, near Burley, and also at Ocknell.

Turtle Dove. (*Columba turtur*, Lin.) Not uncommon. Makes a slight framework of heather for a nest, which it places in a furze bush or low holly. Is extremely shy, and easily forsakes its eggs.

Swift. (*Cypselus apus*, Illig.)

Nightjar. (*Caprimulgus Europæus*, Lin.) Known throughout the Forest as the "Night Hawk," "Night Crow," "Ground Hawk," from its habits, and manner of flying. I have received its eggs at all dates, from the middle of May to the end of July.

Spotted Flycatcher. (*Muscicapa grisola*, Lin.) Arrives about the same time as the three preceding, namely, the beginning of May.

Redbacked Shrike. (*Lanius collurio*, Lin.)

Hobby. (*Falco subbuteo*, Lath.) Generally breeds from the

beginning to the end of June. Mr. Farren, however, in 1861, found a nest containing three eggs so early as May 28th. See Chapter XXII.

Honey Buzzard. (*Falco apivorus*, Lin.) Never arrives before the end of May. See Chapter xxii.

Puffin. (*Mormon fratercula*. Tem.) Comes to the Barton cliffs from the Isle of Wight, where it breeds.

Here, as before, the list clearly indicates the nature of the country. The wheatear proclaims the down-like spaces on the tops of the hills, whilst the hobby and the honey-buzzard tell of the vast extent of woods. In the following division the winter birds speak, instead, of the morasses and bogs, and the river estuaries and mudbanks, which surround the Forest district.

Shorteared owl. (*Strix brachyotus*, Gmel.) Not uncommon. Mr. Cooper, the Forest Keeper to whom I have before referred, tells me that in winter and late in the autumn for twenty years past he has invariably met specimens in heathy and marshy spots at Harvestslade between Burley and Boldrewood. A specimen was killed in November, 1860, in Dibden Bottom, by L. H. Cumberbatch, Esq.

Fieldfare. (*Turdus pilaris*, Lin.) Large numbers frequent the Forest, where it is known as the "blacktail." It especially frequents the hawthorn, and seldom approaches the hollies till the berries of the former are all eaten.

Siskin. (*Fingilla spinus*, Lin.) Now and then taken by the birdcatchers.

Lesser Redpole. (*Fingilla linaria*, Lin.) I should not be surprised if this was discovered to breed in the Forest, as so many pair are seen late in the spring.

Crossbill. (*Loxia curvirostra*, Lin.) Not uncommon. In Dec., 1861, a large flock frequented the plantations round Burley. A few pair are sometimes to be seen in the summer, and Mr. Farren mentions a nest built in a fir-tree in a garden near Lyndhurst,

June, 1858, but the birds were unfortunately not preserved, though their identity is beyond dispute.

Hooded Crow. (*Corvus comix*, Lin.) Not unfrequent.

Golden Plover. (*Charadrius plucialis*, Lin.)

Ringed Plover. (*Charadrius hiaticula*, Lin.) Known, with the dunlin, in the neighbourhood of Christchurch and Lymington, as the "oxbird."

Sanderling. (*Calidris arenaria*, Leach.) Not uncommon on the coast, especially in Christchurch harbour.

Bittern. (*Ardea stellaris*, Lin.) Not a year passes without several specimens being brought to the bird stuffers. Mr. Rake tells me that five were killed close to Fordingbridge in the winter of 1858.

Curlew. (*Numenius arquata*, Lin.)

Green Sandpiper. (*Totanus ochropus*, Tem.) Rather common between Lymington and Calshot Castle. Mr. Rake informs me that a pair were shot at Hale, on the borders of the New Forest, April, 1858; and Mr. Hart tells me that he has shot several in the summer in Stanpit Marsh. In June, 1862, I saw several pair near Leap, so that it probably breeds on the coast.

Jack Snipe. (*Scolopax gallinula*, Lin.) Mr. Cooper tells me that he has known this bird lie so close that he has walked up to it and caught it with his hat.

Knot. (*Tringa Canutus*, Lin.) Not uncommon during the spring at Christchurch Harbour. Mr. Tanner has a specimen in his collection, knocked down with a stick by a boy.

Dunlin. (*Tringa variabilis*, Meyer.) By no means uncommon. See Ringed Plover.

Grey-lag Goose. (*Anser ferus*, Steph.)

Bean Goose. (*Anser segetum*, Gmel.) A stray bird from the Solent sometimes finds its way to Whitten and Ocknell ponds.

Brent Goose. (*Anser bernicla*, Illig.) Locally known as the "Brangoose."

Hooper. (*Cygnus musicus*, Tem.)

Pintail Duck. (*Anas acuta*, Lin.)

Wigeon. (*Anas Penelope*, Lin.)

Common Scoter. (*Anas nigra*, Lin.)

Pochard. (*Anas ferina*, Lin.) Known along the coast as the "redhead" and "ker."

Scaup Duck. (*Anas marila*, Lin.)

Tufted Duck. (*Anas Juligula*, Lin.)

Red-breasted Merganser. (*Mergus serrator*, Lin.) Known to the fishermen at Christchurch as the "razorbill."

Great Crested Grebe. (*Podiceps cristatus*, Lath.) Appears every winter in Christchurch harbour, and may be seen just cresting the waves, as they break under the Barton Cliffs. Mr. Rake informs me that specimens were killed at Breamore, November, 1855, and again, Jan., 1856.

Great Northern Diver. (*Colymbus glacialis*, Lin.)

Red Throated Diver. (*Colymbus septentrionalis*, Lin.) Not so common as the last.

Gannet. (*Sula Bassana*, Boie.)

Blackheaded Gull. (*Larus ridibundus*, Lin.)

Kittiwake. (*Larus tridactylus*, Lin.)

Common Gull. (*Larus canus*, Lin.)

Lesser Blackbacked Gull. (*Larus fuscus*, Lin.) Used formerly to breed in the Freshwater Cliffs of the Isle of Wight.

Great Blackbacked Gull. (*Larus marinus*, Lin.)

The difficulty in the foregoing list has been to decide which species to insert or omit. Many which I have left out, others, perhaps, would have given, will be found placed amongst my last catalogue of stragglers. But before we take these, let me mention two birds of double passage which visit the Forest.

Ring-ousel. (*Turtlus torquatus*, Lin.) A few appear in the spring, but the greater body in the autumn, when they frequent the yews and mountain ashes, being especially fond of the sweet berries of the former. They will hide and skulk, much as a blackbird does, in the furze and brambles, and old thick hedges on the borders of the Forest. Mr. Rake sends me the following interesting note: "An intelligent working man, somewhat, too, of an ornithologist, told me that a few years since he took its nest with four or five eggs, near Ringwood, having a distinct view of the bird as she left the nest."

The Dotterel. (*Charadrius morinellus*, Lin.) Little flocks of them may be seen in the Forest in April, and again in the autumn; but they stay only for a few days.

These are the only two birds which I can satisfactorily class as being truly of double passage. The common sandpiper remains to breed, whilst the grey plover and the whimbrel are killed in the depth of winter. The common redshank, which is generally placed in this division, remains all the year, and the greenshank is seen in the summer, whilst the bar-tailed godwit appears too seldom to admit of being classified in this section. We will therefore go on to the next list, which includes all those birds that cannot be arranged in the foregoing divisions, with the rare stragglers which are driven here by accident, or only appear at uncertain intervals.

Golden Eagle. (*Falco chrysaëlos*, Lin.) The last seen was killed, according to Mr. Hart, about twenty years ago, at the mouth of Christchurch harbour.

Spotted Eagle. (*Falco navius*, Gmel.) A fine male specimen was shot, Dec. 28th, 1861, by a keeper of Lord Normanton's, in the plantations near Somerley. The bird had been noticed for some days previously hovering over the Forest. Mr. Rake, who saw it in the flesh, tells me that the wings measured six feet from tip to tip, and its weight was exactly eight pounds.

White-tailed Eagle. (*Falco albicilla*, Gmel.) *See* Chapter XXII.,

Osprey. (*Falco haliæëtus*, Lin.) Might almost be classed as a regular visitor in the autumn along the coast.

Goshawk. (*Falco palumbarius*, Lin.) Sometimes a stray bird is killed.

Kite. (*Fulco milvus*, Lin.) Very scarce. Mr. Farren, however, in April, 1861, was lucky enough to see a solitary bird; and another, as L. H. Cumberbatch, Esq., informs me, was trapped at New Park, about six years ago, in the winter.

Bough-legged Buzzard. (*Falco lagopus*, Brün.) Mr. Rake informs me that a specimen was trapped near Fordingbridge, in the summer of 1857. It is, however, more generally noticed later in the year.

Little Owl. (*Strix passerina*. Lath.) When Mr. Farren first mentioned this bird as breeding in the Forest, I was somewhat incredulous. Subsequent inquiries, however, have left no doubt on my mind that the bird is sometimes seen, though mistaken for a hawk. Mr. Farren, as far back as 1859, found two eggs in a bole of an oak, which seem to have been those of this bird; and in 1862 I received information of a hawk laying white eggs in a hollow tree, but which were unfortunately broken. I hope, however, some day to be able to give more satisfactory information on the subject.

Ash-coloured Harrier. (*Falco cineraceus*, Mont.) Mr. Hart has, during the last twenty years, received three or four specimens to stuff—one in the winter of 1861. Mr. Farren saw a male bird, April, 1861.

Great Grey Shrike. (*Lanius excubitor*, Lin.) A straggler is now and then killed by the Forest keepers.

Woodchat Shrike. (*Lanius rufus*, Briss.) As some pairs are sometimes to be seen in the summer, I should not be surprised to hear of its breeding, more especially as Mr. Bond has obtained its eggs in the Isle of Wight.

Pied Flycatcher. (*Muscicapa atricapilla*, Lin.) A specimen was shot by the late Mr. Toomer, Forest keeper, June, 1857; but I cannot learn whether male or female.

White's Thrush. (*Turdus Whitei*, Eyton.) Two specimens have been obtained; one in the actual Forest shot by a Forest keeper, and which passed into Mr. Bigge's collection; and the other, not far from its borders at Heron Court, by Lord Malmesbury, and which is figured in Yarrell, vol. i., p. 202. For the best account of this bird see Mr. Tomes' description in the *Ibis*, vol. i., number iv., p. 379, of a specimen killed in Warwickshire.

Golden Oriole. (*Oriolus galbula*, Lin.) A specimen was killed in the Forest by one of the keepers, some fifteen years ago.

Black Redstart. (*Sylvia tithys*, Scop.) I am almost inclined to put this, as Mr. Knox has done in his excellent *Ornithological Rambles* (page 193), and Mr. More in his list of the birds of the Isle of Wight, among the winter visitors, so many examples having occurred.

Great Sedge Warbler. (*Sylvia turdoides*, Meyer.) Mr. Farren, in June, 1858, found between Brockenhurst and Lyndhurst, a nest, containing five eggs, which were supposed to be those of this bird, and were exhibited at a meeting of the Linnæan Society. They are now, I believe, in the collection of Mr. Seeley.

Firecrested Regulus. (*Regulus ignicapillus*, Nawm.) Sometimes seen in the winter, but rare.

Crested Titmouse. (*Parus cristatus*, Lin.) Mr. Hart has once only received a specimen, killed in Stanpit Marsh, near Christchurch. The bird has also been killed in the Isle of Wight.

Bearded Titmouse. (*Parus biarmicus*, Lin.) I once received the eggs of this bird, taken amongst the reeds of the Boldre stream,—the only instance, I believe, of its breeding so far south. The bird has also been seen near Christchurch, among the rushes close to the mouth of the harbour.

Bohemian Waxwing. (*Bombycilla garrula*, Flem.) Mr. Hart tells me that a specimen was shot about twelve years ago at Milton, on the south border of the Forest.

Grayheaded Wagtail. (*Motacilla neglecta*, Gould.) Very rare; but has, on Mr. Hart's authority, been killed.

Short-toed Lark. (*Alauda brachydactyla*, Leisl.) A specimen, caught not far from the Forest boundary, is now in the Rev. J. Bartlett's aviary. See *The Zoologist*, March, 1862, p. 7930.

Snow Bunting. (*Emberiza nicalis*, Lin.) A few are occasionally seen during hard winters.

Brambling. (*Fringilla montifringilla*, Lin.) Occurs like the former bird only during severe frosts. Mr. Rake informs me that a pair were killed near Fordingbridge, in February, 1853.

Tree Sparrow. (*Fringilla montana*, Lin.) Rare.

Mealy Redpole. (*Fringilla borealis*, Tem.) Sometimes caught by the birdcatchers.

Parrot Crossbill. (*Loxia pityopsittacus*, Bechst.) Mr. Rake informs me that one was killed at Breamore, Nov. 28th, 1855, out of a flock of a dozen, and that a few days afterwards several more were killed.

Rose-coloured Pastor. (*Pastor roseus*, Tem.) A fine male was shot some twenty years ago, by Mr. Hart's brother, at Purewell.

Chough. (*Pyrrhocorax graculus*, Tem.) Becoming every year more scarce. *See* Chapter XXII.

Great Black Woodpecker. (*Picus martius*, Lin.) On its breeding habits in Sweden, see Mr. Simpson's account in the *Ibis*, vol. i., p. 264, which agrees about the bird not making a fresh hole, as described at pp. 272, 273.

Hoopoe. (*Upupa epops*, Lin.) *See* Chapter XXII,

White-bellied Swift. (*Cypselus alpinus*, Tem.) Mr. Hart informs me that a specimen was killed about ten years ago over Christchurch harbour.

Rock Dove. (*Columba livia*, Briss.)

Red-legged Partridge. (*Perdix rubra*, Briss) Introduced many years ago by the late Mr. Baring, of Somerley; but very few, if any, are left.

Quail. (*Perdix coturnix*, Lath.) Sometimes to be seen amongst the covies of partridges in the fields adjoining the Forest.

Great Bustard. (*Otis tarda*, Lin._) The last bustard, as

mentioned in Chapter II., p. 14, footnote, was seen about twenty five years ago by one of the Forest keepers, near Eyeworth Wood; but though on horseback, he could not overtake the bird, which ran across Butt's Plain, aiding itself by flapping its wings.

Little Bustard. (*Otis tetrax*, Lin.) A female was shot some years ago near Heron Court; and is in Lord Malmesbury's collection. *See* Eyton's Rarer British Birds, p. 99.

Kildeer Plover. (*Charadrius vociferus*, Lin.) This rare straggler, the only one ever known to have been seen in England, was shot, April, 1859, in a potato field close to Knapp Mill, near Christchurch, by a man of the name of Dowding, who was attracted to it by its peculiar flight, such as is described by Audubon, as also by its monotonous cry, from which its name is taken. The bird was brought in the flesh to Mr. Hart, and is now in the collection of J. Tanner, Esq. The vignette at p. 318 well shows its difference from the common ring dotterel.

Little Ringed Plover. (*Charadrius minor*, Meyer.) Very rare. Mr. Hart has only had one specimen, brought to him many years ago.

Grey Plover. (*Vanellus melanogaster*, Bechst.) Not uncommon during severe winters in the harbours along the coast.

Turnstone. (*Strepsilas interpres*, Ill.) Not uncommon. My friend, Mr. Tanner, has killed both male and female in summer plumage.

Oyster-catcher. (*Hæmatopus ostralegus*, Lin.) By no means uncommon.

Purple Heron. (*Ardea purpurea*, Lin.) One or two specimens have occasionally been shot.

Little Egret. (*Ardea garzetta*, Lin.) Mr. Rake informs me that one was said to have been shot some years ago at Hale, on the borders of the Forest. Yarrell mentions another (vol.ii., p. 554) killed, in 1822, on the Stour near Christchurch.

Squacco Heron. (*Ardea ralloides*, Scop.) A solitary specimen, shot a few years ago at Christchurch Harbour, is now in Lord Malmesbury's collection. See Eyton's *Rarer British Birds*, p. 100, where Dewhurst must probably be a misprint for Christchurch.

Little Bittern. (*Ardea minuta*, Lin.) Mr. Hart, to whom I am under so many obligations for notices of our stragglers, informs me that a fine male bird was shot, April, 26, 1862, on the borders of the Forest, at Heron Court, by one of Lord Malmesbury's keepers.

Night Heron. (*Nycticorax ardeola*, Tem.) Mr. Hart has occasionally received a specimen.

Glossy Ibis. (*Ibis falcinellus*, Tem.) Mr. Hart killed a young pair in a meadow near Christchurch Harbour in September, 1859.

Whimbrel. (*Numenius phæopus*, Lath.) Not so very uncommon during the late autumn and winter mouths along the harbours of the coast.

Spotted Redshank. (*Totanus fuscus*, Leisl.) On the authority of Mr. Hart, who has killed it in Christchurch Harbour.

Avocet. (*Recurvirostra avocetta*, Lin.) Mr. Rake informs me of a specimen shot at Exbury, Dec. 1858.

Blacktailed Godwit, (*Limosa melanura*, Leisl.) Mr. Hart received one in the spring of 1860, and a fine specimen was killed by one of the Forest keepers, some twenty years ago, on Ocknell pond. Hawker, who well knew the sea-coast of the New Forest, mentions large flocks of "grey godwits" off Keyhaven, May, 1842. but he does not distinguish between this and the next species.

Bartailed Godwit. (*Limosa rufa*, Briss.) Mr. Hart had two pair brought to him from the Mudeford Marsh, in the summer of 1861.

Ruff. (*Machetes pugnax*, Cuv.) A specimen is now and then killed.

Great Snipe. (*Scolopax major*, Gmel.) Generally one or two may be seen in the Forest every winter. Mr. Cooper, the Forest keeper, to whom I have previously referred, tells me that during the last twenty years he has shot some six or seven specimens, and has seen as many more killed.

Sabine's Snipe. (*Scolopax Sabini*, Vigors.) *See* Common Snipe (*Scolopax gallinago*), in the list of residents, p. 309.

Curlew Sandpiper. (*Tringa subarquata*. Tem.)

Little Stint. (*Tringa minuta*, Leisl.) Like the preceding, not so very unfrequent along the coast.

Purple Sandpiper. (*Tringa maritima*, Brün.) Occasionally seen in Christchurch Harbour.

Spotted Crake. (*Gallinula porzana*, Lath.) Has been seen both in winter and summer; and I should not be surprised to hear of its breeding.

Baillon's Crake. (*Gallinula Baillonii*, Tem.) A female was shot near Linwood, in the Forest, Nov., 1860.

Grey Phalarope. (*Phalaropus platyrhyncus*, Tem.) Mr. Rake informs me that several specimens were killed on the Avon in the severe winter of 1855-6, and again in 1860-1. Mr. Tanner has a pair in his collection, shot in the mouth of Christchurch Harbour in summer plumage.

Whitefronted Goose. (*Anser albifrons*, Bechst.)

Bernacle Goose. (*Anser leucopsis*, Bechst.) From Mr. Hart I learn that a pair were killed some years ago between Christchurch and Barton.

Ægyptian Goose. (*Anser Ægyptiacus*, Jenyns) From Mr. Rake I learn that a specimen was killed on the Avon, near Bicton Mill, February, 1855.

Bewick's Swan. (*Cygnus minor*, Keys and Bl.)

Shoveller. (*Anas clypeata*, Lin.) Mr. Rake, in his manuscript notes, which he so kindly put in my hands, mentions that this and the gadwall and Bewick's swan, were killed on the Avon during the hard winter of 1855.

Gadwall. (*Anas strepera*, Lin.)

Garganey. (*Anas querquedula*, Lin.)

Eider Duck. (*Anas mollissima*, Lin.)

Velvet Scoter. (*Anas fusca*, Lin.) Sometimes shot by the Mudeford fishermen, but always outside the bar of the harbour.

Long-tailed Duck. (*Anas glacialis*, Lin.)

Golden Eye. (*Anas clangula*, Lin.)

Smew. (*Mergus albellus*, Lin.) Seen, like the two previous, during hard winters on the Avon. Mr. Rake notes that one was killed at Breamore, Nov., 1855; and Mr. Hart writes that he once saw a person kill two at one shot in Christchurch Harbour.

Goosander. (*Mergus merganser*, Lin.) Rather rare. Mr. Rake, however, informs me that one male and two or three females were killed near Fordingbridge in the winter of 1855.

Red-necked Grebe. (*Podiceps ruficollis*, Lath.) Rather rare.

Sclavonian Gbeue. (*Podiceps cornutus*, Lath.) Very rare. Mr. Hart has never known an instance of one being killed, though he has received a specimen or two from the Dorsetshire coast.

Eared Grebe. (*Podiceps auritus*, Lath.) Rather rare, but occasionally killed by the Mudeford fishermen.

Black-throated Diver (*Colymbus arcticus*, Lin.) Occurs pretty plentifully during some winters along the coast.

Little Auk. (*Uria alle*, Tem.) Found sometimes along the coast after a heavy storm.

Caspian Tern. (*Sterna Caspia*, Pall.) On the authority of Mr. Hart one was shot, about ten years ago, in Christchurch Harbour.

Common Tern. (*Sterna hirundo*, Lin.) This, with the next, is sometimes, after a heavy gale, picked up in an exhausted state. I saw one which had been thus caught near Fordingbridge in September, 1861.

Artic Tern. (*Sterna arctica*, Flem.)

Lesser Tern. (*Sterna minuta*, Lin.) Seen during a hard winter.

Black Tern. (*Sterna nigra*, Briss.) A pair were, not long ago, shot by Mr. Charles Reeks, near the Old Bridge, Christchurch.

Little Gull. (*Larus minutus*, Pall.) Mr. Rake informs me that a pair of these rare birds were killed near Breamore, in November, 1855.

Glaucous Gull. (*Larus glaucus*, Brün.) A solitary specimen has, I believe, once been shot near Christchurch, by the Hon. Grantley Berkeley, in whose collection it is.

Common Skua. (*Lestris catarractes*, Ill.) Occasionally killed flying round Christchurch Head.

Fork-tailed Petrel. (*Thalassidroma Leachii*, Tem.) Mr. Rake informs me that a specimen was picked up dead, near Fordingbridge, November, 1859.

Stormy Petrel. (*Thalassidroma pelagica*, Vigors.) Frequently picked up dead, or exhausted, along the coast, after severe weather, with the wind blowing from the west.

Adopting Yarrell's census, an analysis of these lists gives to the Forest district 72 out of the 140 British residents, 31 out of our 63 summer visitors, 35 winter visitors, and of rarer birds and stragglers, 90; or altogether, including the two birds of double passage, 230 species out of the whole 354.

Since these lists were printed, Mr. Rake sends me word, concerning the reed wren, that in the winter of 1858, a nest, evidently built the preceding summer, and exactly resembling that bird's, was found in a thick bed of reeds on the bank of the Avon, near Fordingbridge, but he has never seen the birds or eggs from the neighbourhood.

With regard to the kildeer plover, I may add that several persons saw it in the flesh, and that Mr. Tanner received it soon after it was mounted. My only surprise is with Dr. Sclater (see the *Ibis* vol. iv., No. xv., p. 277), that a bird with so large a range of flight should not before this have been recorded as occurring in England.

The vignette is, with a slight alteration of position, taken from Mr. Tanner's specimen.

The Kildeer Plover

APPENDIX IV.

THE LEPIDOPTERA OF THE NEW FOREST.

As, I am sorry to say, I am entirely ignorant of entomology, Mr. Baker, who possesses one of the finest collections of Lepidoptera in the district, has kindly compiled the following list. For the sake of space the Tineina have been omitted. The arrangement followed is that of Stainton, and the whole list has, to ensure the greatest accuracy, been revised by F. Bond, Esq., F.Z.S. No attempt has been made to classify the rarer and more common species, as both so much vary with the season.

RHOPALOCERA.

GONOPTERYX
Rhamni
COLIAS
Edusa
Hyale
APORIA
Cratægi
PIERIS
Brassicæ
Rapæ
Napi
ANTHOCHARIS
Cardamines
LEUCOPHASIA
Sinapis
ARGE
Galathea
LASIOMMATA
Ægeria
Megæra
HIPPARCHIA
Semele
Janira
Tithonus
Hyperanthus
CŒNONYMPHA
Pamphilus
LIMENITIS
Sibylla
APATURA
Iris
VANESSA
Cardui
Atalanta
Io
Antiopa
Polychloros
Urticæ
ARGYNNIS
Paphia
Adippe
Aglaia
Selene
Euphrosyne
MELITÆA
Artemis
NEMEOBIUS
Lucina
THECLA
Betulæ
Quercus
Rubi
CHRYSOPHANUS
Phlœas
POLYOMMATUS
Argiolus
Alsus
Alexis
Ægon
Agestis
THYMELE
Alveolus
THANAOS
Tages
STEROPES
Paniscus
PAMPHILA
Linea
Sylvanus

319

SPHINGINA.

- PROCRIS
  - Statices
- ANTHROCERA
  - Trifolii
  - Loniceræ
  - Filipendulæ
- SMERINTHUS
  - Ocellatus
  - Populi
  - Tiliæ
- ACHERONTIA
  - Atropos
- SPHINX
  - Convolvuli
  - Ligustri
- DEILEPHILA
  - Galii
- CHŒROCAMPA
  - Elpenor
  - Porcellus
- MACROGLOSSA
  - Stellatarum
- SESIA
  - Fuciformis
  - Bombyliformis
- SPHECIA
  - Bembeciformis
- TROCHILIUM
  - Ichneumoniforme
  - Cynipiforme
  - Sphegiforme
  - Tipuliforme
  - Myopæforme

BOMBYCINA.

- HEPIALUS
  - Hectus
  - Lupulinus
  - Humuli
  - Sylvinus
- ZENZERA
  - Æsculi
- COSSUS
  - Ligniperda
- CERURA
  - Furcula
  - Vinula
- STAUROPUS
  - Fagi
- NOTODONTA
  - Dromedarius
- DRYMONIA
  - Chaonia
  - Dodonæa
- LEIOCAMPA
  - Dictæa
  - Dictæoides
- LOPHOPTERYX
  - Camelina
- DILOBA
  - Cæruleocephala
- PETASIA
  - Cassinia
- PERIDEA
  - Trepida
- CLOSTERA
  - Reclusa
- PYGÆRA
  - Bucephala
- PSILURA
  - Monacha
- DASYCHIRA
  - Fascelina
  - Pudibunda
- DEMAS
  - Coryli
- ORGYIA
  - Antiqua
- STILPNOTIA
  - Salicis
- PORTHESIA
  - Auriflua
- MILTOCHRISTA
  - Miniata
- LITHOSIA
  - Aureola
  - Helvola
  - Stramineola
  - Complana
  - Complanula
  - Griseola
- ŒNISTIS
  - Quadra
- GNOPHRIA
  - Rubricollis
- CYBOSIA
  - Mesomella
- NUDARIA
  - Mundana
  - Senex
- EUTHEMONIA
  - Russula
- ARCTIA
  - Caja
  - Villica
- NEMEOPHILA
  - Plantaginis
- PHRAGMATOBIA
  - Fuliginosa
- SPILOSOMA
  - Menthastri
  - Lubricipeda
- DIAPHORA
  - Mendica
- CALLIMORPHA
  - Jacobææ
- EULEPIA
  - Cribrum
- DEIOPEIA
  - Pulchella
- LASIOCAMPA
  - Rubi
  - Trifolii
  - Quercus
- ERIOGASTER
  - Lanestris
- PŒCILOCAMPA
  - Populi
- TRICHIURA
  - Cratægi
- CLISIOCAMPA
  - Neustria
- ODONESTIS
  - Potatoria

BOMBYCINA.—(*Continued.*)

GASTROPACHA
Quercifolia
SATURNIA
Pavonia-minor
CILIX
Spinula
PLATYPTERYX
Lacertinaria
DREPANA
Falcataria
Hamula
Unguicula
HETEROGENEA
Asellus
LIMACODES
Testudo

PSYCHIDÆ.

PSYCHE
Nigricans
Opacella
FUMEA
Radiella

NOCTUINA.

THYATIRA
Derasa
Batis
CYMATOPHORA
Duplaris
Diluta
Flavicornis
Ridens
BRYOPHILA
Perla
DIPHTHERA
Orion
ACRONYCTA
Tridens
Psi
Leporina
Megacephala
Alni
Ligustri
Rumicis
LEUCANIA
Conigera
Turca
Lithargyria
Pudorina
Comma
Impura
Pallens
NONAGRIA
Despecta
Fulva
Typhæ
GORTYNA
Flavago
HYDRŒCIA
Nictitans
Micacea
AXYLIA
Putris
XYLOPHASIA
Rurea
Lithoxylea
Polyodon
Hepatica
DIPTERYGIA
Pinastri
NEURIA
Saponariæ
HELIOPHOBUS
Popularis
CHARÆAS
Graminis
CERIGO
Cytherea
LUPERINA
Testacea
Cæspitis
MAMESTRA
Anceps
Furva
Brassicæ
Persicariæ
APAMEA
Basilinea
Gemina
Oculea
MIANA
Strigilis
Fasciuncula
Literosa
Furuncula
Arcuosa
CELÆNA
Haworthii
GRAMMESIA
Trilinea
ACOSMETIA
Caliginosa
CARADRINA
Morpheus
Alsines
Blanda
Cubicularis
RUSINA
Tenebrosa
AGROTIS
Puta
Suffusa
Saucia
Segetum
Exclamationis
Nigricans
Tritici
Aquilina
Porphyrea
Ravida
TRYPHÆNA
Ianthina
Fimbria
Interjecta
Subsequa
Orbona
Pronuba
NOCTUA
Glareosa
Augur
Plecta
C-nigrum

T T

NOCTUINA.—(*Continued.*)

Triangulum
Brunnea
Festiva
Bella
Umbrosa
Baja
Neglecta
Xanthographa
TRACHEA
Piniperda
TÆNIOCAMPA
Gothica
Rubricosa
Instabilis
Stabilis
Gracilis
Miniosa
Munda
Cruda
ORTHOSIA
Upsilon
Lota
Macilenta
ANCHOCELIS
Rufina
Pistacina
Lunosa
Litura
CERASTIS
Vaccinii
Spadicea
SCOPELOSOMA
Satellitia
DASYCAMPA
Rubiginea
OPORINA
Croceago
XANTHIA
Citrago
Cerago
Silago
Aurago
Gilvago
Ferruginea
TETHEA
Subtusa
Retusa
DICYCLA
Oo
COSMIA
Trapezina
Pyralina
Diffinis
Affinis
EREMOBIA
Ochroleuca
DIANTHŒCIA
Carpophaga
Capsincola
Cucubali
HECATERA
Serena
POLIA
Flavicincta
EPUNDA
Lutulenta
Nigra
Viminalis
MISELIA
Oxyacanthæ
AGRIOPIS
Aprilina
PHLOGOPHORA
Meticulosa
EUPLEXIA
Lucipara
APLECTA
Herbida
Nebulosa
Advena
HADENA
Adusta
Protea
Dentina
Chenopodii
Suasa
Oleracea
Pisi
Thalassina
Contigua
Genistæ
XYLOCAMPA
Lithorhiza
CALOCAMPA
Vetusta
Exoleta
XYLINA
Rhizolitha
Semibrunnea
Petrificata
CUCULLIA
Chamomillæ
Umbratica
HELIOTHIS
Marginata
Dipsacea
ANARTA
Myrtilli
HELIODES
Arbuti
ACONTIA
Luctuosa
ERASTRIA
Fuscula
HYDRELIA
Uncana
BREPHOS
Parthenias
HABROSTOLA
Urticæ
Triplasia
PLUSIA
Chrysitis
Iota
Pulchrina
Gamma
GONOPTERA
Libatrix
AMPHIPYRA
Pyramidea
Tragopogonis
MANIA
Typica
Maura
TOXOCAMPA
Pastinum
STILBIA
Anomala
CATOCALA
Nupta
Promissa
Sponsa
EUCLIDIA
Mi
Glyphica
PHYTOMETRA
Ænea

GEOMETRINA.

URAPTERYX
Sambucaria
EPIONE
Apiciaria
Advenaria
RUMIA
Cratægata
VENILIA
Maculata
ANGERONA
Prunaria
METROCAMPA
Margaritaria
ELLOPIA
Fasciaria
URYMENE
Dolabraria
ERICALLIA
Syringaria
SELENIA
Illunaria
Lunaria
Illustraria
ODONTOPERA
Bidentata
CROCALLIS
Elinguaria
ENNOMOS
Tiliaria
Fuscantaria
Erosaria
Angularia
HIMERA
Pennaria
PHIGALIA
Pilosaria
NYSSIA
Hispidaria
AMPHIDASYS
Prodromaria
Betularia
HEMEROPHILA
Abruptaria
CLEORA
Viduaria
Glabraria
Lichenaria
BOARMIA
Repandaria
Rhomboidaria
Abietaria
Cinctaria
Roboraria
Consortaria
TEPHROSIA
Consonaria
Crepuscularia
Extersaria
Punctularia
GNOPHOS
Obscurata
Pullata
PSEUDOTERPNA
Cytisaria
GEOMETRA
Papilionaria
MEMORIA
Viridata
IODIS
Lactearia
PHORODESMA
Bajularia
HEMITHEA
Thymiaria
EPHYRA
Poraria
Punctaria
Trilinearia
Omicronaria
Orbicularia
Pendularia
HYRIA.
Auroraria
ASTHENA
Luteata
Candidata
Sylvata
EUPISTERIA
Heparata
ACIDALIA
Scutulata
Bisetata
Trigeminata
Osseata
Virgularia
Ornata
Incanaria
Marginepunctata
Subsericeata
Immutata
Remutata
Imitaria
Aversata
Emarginata
BRADYEPETES
Amataria
CABERA
Pusaria
Exanthemaria
CORYCIA
Temerata
Taminata
AVENTIA
Flexula
MACARIA
Alternata
Notata
Liturata
HALIA
Vauaria
STRENIA
Clathrata
LOZOGRAMMA
Petraria
NUMERIA
Pulveraria
MÆSIA
Belgiaria
SELIDOSEMA
Plumaria
FIDONIA
Atomaria
Piniaria
MINOA
Euphorbiata

T T 2

GEOMETRINA.—(*Continued.*)

ASPILATES
Strigillaria
ABRAXAS
Grossulariata
LIGDIA
Adustata
LOMASPILIS
Marginata
PACHYCNEMIA
Hippocastanaria
HYBERNIA
Rupicapraria
Leucophæaria
Aurantiaria
Progemmaria
Defoliaria
ANISOPTERYX
Æscularia
CHEIMATOBIA
Brumata
OPORABIA
Dilutata
LARENTIA
Didymata
Multistrigaria
Pectinitaria
EMMELESIA
Affinitata
Alchemillata
Albulata
Decolorata
Unifasciata
EUPITHECIA
Venosata
Linariata
Pulchellata
Centaureata
Succenturiata
Subumbrata
Haworthiata
Pygmæata
Satyrata
Castigata
Irriguata
Denotata
Innotata
Indigata
Nanata
Subnotata
Vulgata
Expallidata
Absinthiata
Minutata
Assimilata
Tenuiata
Dodoneata
Abbreviata
Exiguata
Pumilata
Coronata
Rectangulata
LOBOPHORA
Sexalata
Hexapterata
Viretata
Lobulata
THERA
Variata, *Haw.*
Firmaria
HYPSIPETES
Impluviata
Elutata
MELANTHIA
Rubiginata
Ocellata
Albicillata
MELANIPPE
Unangulata
Rivata
Subtristrata
Montanata
Fluctuata
ANTICLEA
Rubidata
Badiata
Derivata
COREMIA
Propugnata
Ferrugata
CAMPTOGRAMMA
Bilineata
Gemmata
SCOTOSIA
Dubitata
Certata
Undulata
CIDARIA
Psittacata
Miata
Picata
Corylata
Russata
Immanata
Suffumata
Silaceata
Prunata
Testata
Fulvata
Pyraliata
Dotata
EUBOLIA
Cervinaria
Mensuraria
Palumbaria
Bipunctaria
Lineolata
ANAITIS
Plagiata
ODEZIA
Chærophyllata

PYRALIDINA.

DELTOIDES.

HYPENA
- Proboscidalis
- Rostralis

HYPENODES
- Costæstrigalis

RIVULA
- Sericealis

HERMINIA
- Barbalis
- Tarsipennalis
- Nemoralis

PYRALITES

PYRALIS
- Costalis
- Farinalis
- Glaucinalis

AGLOSSA
- Pinguinalis

CLEDEOBIA
- Angustalis

PYRAUSTA
- Punicealis
- Purpuralis
- Ostrinalis

HERBULA
- Cæspitalis

ENNYCHIA
- Cingulalis
- Octomaculalis

ENDOTRICHA
- Flammealis

DIASEMIA
- Literalis

CATACLYSTA
- Lemnalis

PARAPONY
- Stratiotalis

HYDROCAMPA
- Nymphæalis
- Stagnalis

BOTYS
- Pandalis
- Verticalis
- Lancealis
- Fuscalis
- Urticalis

EBULEA
- Crocealis
- Sambucalis

PIONEA
- Forficalis

SPILODES
- Sticticalis
- Cinctalis

SCOPULA
- Olivalis
- Prunalis
- Ferrugalis

STENOPTERYX
- Hybridalis

NOLA
- Cucullatella
- Cristulalis
- Strigula

SYMAETHIS
- Fabriciana

CHOREUTES
- Scintillulana

CRAMBITES.

EUDOREA
- Cembræ
- Ambigualis
- Pyralella
- Cratægella
- Frequentella
- Resinea
- Pallida

APHOMIA
- Colonella

ACHROIA
- Grisella

EPHESTIA
- Elutella

HOMÆOSOMA
- Nebulella

ACROBASIS
- Consociella
- Tumidella

CRYPTOBLAB
- Bistriga

MYELOIS
- Suavella
- Advenella
- Marmorea

NEPHOPTERY
- Abietella
- Roborella

PEMPELIA
- Dilutella
- Formosa
- Palumbella

CRAMBUS
- Cerussellus
- Chrysonychellus
- Falsellus
- Pratellus
- Dumetellus
- Sylvellus
- Hamellus
- Pascuellus
- Uliginosellus
- Hortuellus
- Culmellus
- Inquinatellus
- Geniculeus
- Contaminellus
- Tristellus
- Pinetellus
- Latistrius
- Perlellus

CHILO
- Forficellus
- Phragmitellus

TORTRICINA.

CHLOEPHORA
Prasinana
Quercana
SAROTHRIPA
Revayana
HYPERMECIA
Angustana
EULIA
Ministrana
BRACHYTÆNIA
Semifasciana
ANTITHESIA
Betuletana
Ochroleucana
Cynosbatella
Pruniana
Marginana
Similana
Sellana
PENTHINA
Salicella
SIDERIA
Achatana
DICHELIA
Grotiana
CLEPSIS
Rusticana
TORTRIX
Icterana
Viburnana
Forsterana
Heparana
Ribeana
Cinnamomeana
Corylana
LOZOTÆNIA
Sorbiana
Musculana
Costana
Unifasciana
Fulvana
Roborana
Xylosteana
Rosana
DITULA
Angustiorana
PTYCHOLOMA
Lecheana
NOTOCELIA
Uddmanniana
PARDIA
Tripunctana
SPILONOTA
Roborana
Trimaculana
Amœnana
LITHOGRAPHIA
Campoliliana
Nisella
Penkleriana
PHLÆODES
Tetraquetrana
Immundana
PŒDISCA
Piceana
Solandriana
Opthalmicana (?)
CATOPTRIA
Scopoliana
Hohenwarthiana
HALONOTA
Bimaculana
Cirsiana
Scutulana
Brunnichiana
DICRORAMPHA
Petiverella
Sequana
Politana
Plumbagana
Consortana
COCCYX
Hercyniana
CAPUA
Ochraceana
CARTELLA
Bilunana
HEDYA
Paykulliana
Ocellana
Dealbana
Trimaculana
STEGANOPTYCHA
Nævana
ANCHYLOPERA
Mitterpacherana
Subarcuana
Biarcuana
Uncana
Lundana
Derasana
Comptana
Siculana
BACTRA
Lanceolana
Furfurana
ARGYROTOXA
Conwayana
DICTYOPTERYX
Contaminana
Lœflingiana
CRŒSIA
Bergmanniana
Forskaleana
Holmiana
HEMEROSIA
Rheediella
OXYGRAPHA
Literana
PERONEA
Schalleriana
Comparana
Tristana
Rufana
Favillaceana
Hastiana
Cristana
Variegana
PARAMESIA
Aspersana
Ferrugana
TERAS
Caudana
PŒCILOCHROMA
Profundana
Corticana
ANISOTÆNIA
Ulmana

TORTRICINA.—(*Continuea.*)

ROXANA
Arcuella
SEMASIA
Populana
Spiniana
Wœberiana
Janthinana
EUCELIS
Aurana
EPHIPPIPHORA
Trauniana
Regiana
Argyrana
STIGMONOTA
Nitidana
Wierana
Compositella
Perlepidana
ASTHENIA
Splendidulana
RETINIA
Buoliana
Pinivorana
Sylvestrana
ENDOPISA
Ulicana
Germarana
Puncticostana
CARPOCAPSA
Juliana
Splendana
Pomonana
GRAPHOLITHA
Albersana
Hypericana
Modestana
SPHALEROPTERA
Ictericana
CNEPHASIA
Hybridana
Subjectana
Passivana
Nubilana
EUCHROMIA
Ericetana
Striana
SERICORIS
Conchana
Lacunana
Urticana
Cæspitana
Politana
Latifasciana
Bifasciana
MIXODIA
Schulziana
LOBESIA
Reliquana
PHTHEOCROA
Rugosana
ERIOPSELA
Fractifasciana
CHROSIS
Tesserana
ARGYROLEPIA
Æneana (?)
Baumanniana
Badiana
CALOSETIA
Nigromaculana
EUPŒCILIA
Maculosana
Carduana
Nana
Angustana
Griseana
Roseana
Subroseana
Ruficiliana
LOZOPERA
Francillana
Stramineana
XANTHOSETIA
Hamana
Zœgana
TORTRICODES
Hyemana

PTEROPHORINA.

PTEROPHORUS
Trigonodactylus
Acanthodactylus
Punctidactylus
Bipunctidactylus
Fuscus
Pterodactylus
Tephradactylus
Galactodactylus
Tetradactylus
Pentadactylus
ALUCITA
Polydactyla

POSTSCRIPT.—As a further addition to my list of plants, I have just received the following from A. G. More, Esq., F.L.S.—those without localities being communicated to him by the late Mr. Borrer as found in the Forest:—

WAHLENBERGIA HEDERACEA, Reich., Ivy-leaved Bell Flower. Near Lyndhurst, 272.

SIUM LATIFOLIUM, Lin., Broad-leaved Water Parsnep. *See* Bromfield, in the *Phytologist*, vol. iii. p. 403; 464.

TRIFOLIUM MEDIUM, Lin., Zigzag Trefoil. Near Lyndhurst. 683.

UTRICULARIA INTERMEDIA, Hayne, Intermediate Bladder-wort, 876.

CAREX LIMOSA, Lin., Green-and-Gold Carex, 1244.

A word or two may here be added concerning the only true species of cicada (*Cicada hæmatoides*), which we have in England, and which has hitherto been only found in the New Forest. Mr. Farren, in June, 1858, was fortunate enough to take a specimen sitting on the stem of the common brake, being attracted to it by its peculiar monotonous humming noise. On the second of June, 1862, he captured two others, which rose from the fern, with their curious zigzag flight, and at the same time heard two more.

Mr. Farren, to whom I am indebted for the above information, has kindly sent me the following drawing, made by his brother, from one of the living specimens captured in the Forest.

The Cicada.

Printed in July 2019
by Rotomail Italia S.p.A., Vignate (MI) - Italy